Natural Capital and Human Economic Survival

Second Edition

ECOLOGICAL ECONOMICS SERIES
Robert Costanza *Series editor*

—Foreword by
Paul Hawken

Natural Capital and Human Economic Survival

Second Edition

by
Thomas Prugh

with
Robert Costanza, John H. Cumberland
Herman E. Daly, Robert Goodland,
and Richard B. Norgaard

ISEE
International
Society for
Ecological
Economics

LEWIS PUBLISHERS
Boca Raton London New York Washington, D.C.

Project Editor: Sara Rose Seltzer
Marketing Managers: Barbara Glunn, Jane Stark,
                         Jane Lewis, Arline Massey
Cover Design: Dawn Boyd

Library of Congress Cataloging-in-Publication Data

Prugh, Thomas.
    Natural capital and human economic survival / by Thomas Prugh ;
with Robert Costanza ... [et al.]. -- 2nd ed.
            p. cm
    Includes bibliographical references and index.
    ISBN 1-56670-398-0 (alk. paper)
    1. Environmental economics. 2. Environmental policy--Economic
aspects. I. Costanza, Robert. II. Title.
    HC79.E5P748   1999
    333.7--DC21

                                                      99-12121
                                                           CIP

© 1999 by CRC Press
Lewis Publishers is an imprint of CRC Press LLC

No claim to original U.S. Government works
International Standard Book Number 1-56670-398-0
Library of Congress Card Number 99-12121
Printed in the United States of America 1 2 3 4 5 6 7 8 9 0
Printed on acid-free paper

For Evanne and Gordon

Nature! We are surrounded and embraced by her: powerless to separate ourselves from her, and powerless to probe beyond her.

— Johann von Goethe
Essays on the Lessons of Nature and Science

It is perfectly clear...that we are indissolubly one with nature and depend completely on the natural environment. Anybody can do a simple experiment to find out how much he depends on the natural environment even though he lives in a world of television and automobiles. He merely has to put a clothespin on his nose and tape up his mouth to find out that he can't do without his natural environment for more than about sixty seconds.

— Aldous Huxley
The Human Situation

Contents

Section III: Managing Natural Capital for Sustainability

Foreword

Great ideas, in hindsight, seem obvious. The concept of natural capital is such an idea: the proposition that no organized economic system can long endure without taking the flow of renewable and nonrenewable resources into account. This revision of neoclassical economics, yet to be accepted by mainstream academicians, stabilizes both the theory and practice of free-market capitalism. It is as if we had been sitting on a two-legged stool for the past century, wondering why our economies are increasingly unbalanced and unsteady. The concept of natural capital, when it is intelligently linked to the concepts of human and manufactured capital, provides the critical nexus between the satisfaction of human needs and the preservation — some might even suggest the restoration — of our living natural systems.

At first glance, this book may seem but one more proof that economics remains the dismal science. Rather than reminding us of our baser instincts, it points to yet another limiting factor: that living systems cannot be replaced by manufactured or human endeavor. In America we have been surfeited with cornucopian fantasies of technological prowess wherein human ingenuity bypasses natural limits and creates unimagined abundance. Optimism easily intertwines with the belief that nanotechnology, biotechnology, computers, and technologies yet to be developed will eliminate hunger, disease, and want. Dreams of alleviating human suffering are worthy, but what they usually overlook is fertile soil, clean air, biological diversity, and pure water, all of which we are rapidly losing, none of which can be substantively created by any man-made technology known or imagined. In our pursuit of dominance over the natural world, we disregarded the basic principles of how life is created and maintained on this nonlinear, self-organizing system we call Earth.

Natural Capital and Human Economic Survival is a lucid synthesis that demonstrates that we cannot have a healthy economy without integrating the dynamics and principles of living systems into our business and politics. To call for the incorporation of these principles into economic theory is not a limiting factor; it is a guide, a template, a pointer to the future. Unless we integrate natural capital formation into our economic theories and commercial systems, we will continue to witness the deteriorating effects of our denial in both social and environmental degradation.

The fact that we have excluded natural capital from the balance sheet of companies, countries, and the world is an extraordinary omission and error, and it has led to an equally grave error with respect to work. We have spent the better part of the past 200 years trying to make human resources more productive: an admirable goal. But to do this, we have employed more and more natural capital to generate energy, metals, roads, buildings, factories, infrastructure, and worldwide communication systems. When resources were abundant, this was an acceptable trade-off. Now that we are 100 times more productive than we were at the beginning of the 19th century and life-sustaining resources are rapidly depleting, this exchange no longer makes sense.

The irony is that not only have we caused great resource depletion by pursuing labor productivity gains, but we are now depleting our people as well. Just as overproduction can exhaust topsoil, overproduction can exhaust a workforce. The underlying assumption that greater productivity would lead to greater leisure and well-being, while true for many decades, has now become a bad joke. In the United States, those who are employed, and presumably becoming more productive, find themselves working 100–200 hours more per year now than 20 years ago. Yet real incomes in the U.S. have not exceeded 1973 levels. Those whose jobs have been eliminated, downsized, re-engineered, and restructured observe this folly with less charity. They are being told, in so many words, as are millions of youth around the world, that we have created an economic system so ingenious and clever that they are not needed, unless it is to fill menial service jobs. This productivity trap is the end result of all linear systems: exhaustion and depletion.

As Thomas Prugh and his co-authors have made abundantly clear herein, there are only two solutions to our continuous assault on our resource base: one, we must slow and stabilize population growth and thereby the demands made upon living systems by the human population, and two, we must discover ways to dramatically improve the productivity of natural capital. Population issues are being worked on by agencies around the world. The effort to reduce population growth is being led by women who pay a high price when resources cannot support their bodies and families. The second task, to increase resource productivity and consequently reduce the impact each person has on the environment, may not be as difficult as it sounds. The reason is that industrialism, for all its sophistication, is enormously inefficient with respect to resources, energy, and waste.

Industrialism had its roots in the appropriation by Europeans of global resources that they thought were theirs: those that they had "discovered." The possession of those resources and their transformation into goods and services through the European production system characterized, and still characterizes, all industrial systems, including the so-called information age. Charitably speaking, we can see how difficult it is for neoclassical economists, whose theories originated during a time of resource abundance, to understand that the very success of linear industrial systems has laid the groundwork for the next stage in economic evolution. This next stage, whatever it may be called, is being brought about by powerful and much-delayed feedback. Information from destructive activities going back 100 years or more is being incorporated into the market and political economy. As that happens, the foundation of industrialism is crumbling. The industrialized world no longer provides workers with the amenities and the "better life" promised. We are surrendering our living systems, social stability, fiscal soundness, and personal health to outmoded economic assumptions. We are hoping that conventional economic growth will save us.

Nationally and globally we still perceive the problems of social and environmental decay as distinct and unconnected, rather than as inextricably rooted in linear industrial systems. When stripped of its advertisements, golf courses, and "affluenza," industrialism simply does not work very well. If its massive inefficiencies are not more apparent, it is because they are masked by a financial system that supplies improper information. This is a classic case of "garbage in, garbage out." The "garbage in" is what money tells us, what prices tell us, what the markets tell us. Instead of markets giving us proper information about how much our suburbs, spandex, and plastic spring water bottles truly cost, everything else is giving us proper information — our beleaguered air and our threatened watersheds, the overworked soils, the deracinated inner cities and rural counties, the breakdown of stability worldwide, the conflicts based on resource shortages; all these are providing the information that our prices should be giving us but don't.

Our prices have failed us for the most simple and frustrating of reasons: bad accounting. Natural capital has never been placed on the balance sheets of companies, countries, or the world. Paraphrasing G. K. Chesterton, it could be fairly said that capitalism might be a good idea except that we have never tried it yet. And try it we must and will, for capitalism cannot be fully attained or practiced until, as any accounting student will tell us, we have an accurate balance statement. As it stands, our economic system is based on accounting principles that would bankrupt a company. Not surprisingly, it is posing problems for the world as a whole. When natural capital is placed on the balance sheet, not as a free amenity of infinite supply, but as an integral and valuable part of the production process, everything changes. The near maniac pursuit of human productivity becomes far less important than the increase of natural resource productivity. Using more and more resources to make fewer people more productive flies in the face of what we now need and require to improve our society and environment. The promise of economic systems based on natural capital is that it can change this century's sole emphasis on human productivity and begin improving natural resource productivity with the attendant benefit of employing more people in work that is meaningful and dignified. After all, it is people we have more of, not natural resources, so it is people we must use more of in order to reduce the throughput of materials and energy from the environment. Students of this book can imagine a world of many people who, rather than being marginalized by technology and concentrated forms of production, find they are needed to reverse this long history of resource abuse and overuse.

For many, the prospect of an economic system based on natural capital productivity is insufficiently dismal, thus calling into question its economic viability. To answer that question, we may well want to reverse the question and ask how it is that we have created an economic system that tells us it is cheaper to destroy the Earth than it is to maintain it. We know this is not true of our cars, houses, and bridges, but somehow we have managed to overlook a pricing system that discounts the future and sells off the past. Or to put it another way, how did we create an economic system that confuses capital with income? (In a publicly held company, such confusion is called fraud, and would come with a jail sentence.) Are we so far from a rational economic system? I think not. It is right before us. It requires no new theories, only common sense. It is based on the proposition that all capital be valued. While there may be no "right" way to value a forest or a river, there is a wrong way, which is to give it no value at all. If we have doubts how to value a 700-year-old tree, we need only ask how much it would cost to make a new one. Or a new river, or even a new atmosphere. The work of the future is the absorption and integration of the worth of natural and living systems into every aspect of our culture and practices, so that human systems are mimetic of natural systems, and so that our culture reflects growth and harmony rather than suffering and discord. Rilke wrote that "to work with things in the indescribable relationship is not too hard for us; the pattern grows more intricate and subtle, but being swept along is not enough. Take your practiced powers and stretch them out until they span the chasm between the two contradictions...." That is the promise of this book, and it is the promise of this work.

Paul Hawken

Preface

This book examines the relationship between human economic activity and environmental degradation from the viewpoint of the transdisciplinary field of ecological economics. Since the "eco" in both ecology and economics comes from the Greek word *oikos* (house), it might be said that this book is about home maintenance. The Earth was once a cavernous mansion where the human family, relatively few in number, lived for many millennia without causing too much harm or running short of space or supplies. Now we have occupied every available bedroom and have begun crowding into the less congenial niches, even the cellar and the broom closets. The floors are sagging, the cisterns sometimes run dry, and the lights flicker now and then. The place needs some work.

Natural Capital and Human Economic Survival is not a detailed argument in support of the claims of ecological crisis. Those claims are well documented in some cases, less so in others. People react to them with concern, alarmism, denial, or indifference, as they are inclined. There is vast literature on the subject and the authors have only touched its surface in Section II.

Our own belief is that the available evidence of ecological danger is profoundly disturbing. We also think there is no doubt that the potential consequences of runaway environmental transformation are frightening enough that we owe it to ourselves, our descendants, and all other forms of life on the planet to be prudent. Because natural capital is the foundation upon which our ecological/economic house rests, there is nothing more prudent than ensuring its continuing soundness. We therefore attempt to review the conceptual and practical problems involved and suggest ways to begin the reconstruction. This book reflects the belief that a synthesis of ecological ideas and economic theory is necessary to correct the historical errors economics has made concerning humanity's relationship to the ecosphere. Because of the power of economists and economic ideas, that is a prerequisite to achieving truly sustainable development.

By design, this book draws upon two other volumes sponsored by the International Society for Ecological Economics, as well as many other sources. The two ISEE books are *An Introduction to Ecological Economics* (St. Lucie Press, 1997), which is a more detailed and analytical treatment of the development of ecological economics and its principles intended for advanced readers and graduate courses, and *Investing in Natural Capital: The Ecological Economics Approach to Sustainability* (Island Press, 1994), a technical volume based on papers prepared for an ISEE workshop in Stockholm, Sweden, August 3-6, 1992. Sections of this book that draw extensively from these sources are so noted. All sources can be found in the References.

We ask readers to bear in mind that ecological economics is not an academic discipline but a transdisciplinary way of viewing environmental problems. These problems, among the most critical of our time, are complex and often poorly understood, so there are many viewpoints and voices. Ecological economics is a large tent that tries to make room for them. This short work cannot hope to do proper justice to all or to reconcile the many differences of opinion held by sincere people. We have tried to distill the essence of ecological economic thinking into a few pages, and hope readers will understand if the material sometimes seems to resist integration. There is still much to be learned about how all the pieces of the puzzle fit together and what sort of picture they present.

What This Book Is About

The central ideas in this book are the following:

1. The global ecosystem (the natural environment) provides a vast array of indispensable resources and services to human beings. Viewed this way, the environment is a form of capital, here called "natural capital." Natural capital is necessary for human economic activity and survival. Natural capital can never be entirely replaced by any combination of human labor, wealth, and technology, although it is sometimes implied or assumed otherwise.

2. The Earth's natural capital endowment is under severe strains from rapidly increasing human economic activity and population. The available evidence of environmental problems strongly suggests that many forms of natural capital are, or are becoming, seriously degraded and scarce. If this continues, the remaining natural capital will be inadequate to allow human beings to continue living on the planet as we have in the past, much less to expand the global economy as nearly

everyone demands. Accordingly, conserving and investing in natural capital should be among our most urgent priorities.

3. The view of natural capital offered by mainstream (neoclassical) economics suffers from a grave theoretical flaw that has critical policy implications. Mainstream economics views natural capital as only a single, rather unimportant, factor of economic production. In contrast, ecological economics views it as the very foundation of the economy. Without natural capital, human economic activity is impossible. Although neoclassical economics did not create the tendency to over-exploit natural capital, it has aided and abetted that overexploitation and blinded citizens and policymakers to the need for a restructuring of the economic rules we play by.

4. Revenue-neutral policy options exist that could restructure the economic system so as to encourage conservation of, and investment in, natural capital. Such policies would help ensure our economic viability into the indefinite future. They would work, in part, by properly valuing natural capital resources and services, thus accounting for the indispensable contribution natural capital makes to economic production. By helping to "get the prices right," they would promote true economic efficiency. This is the first and most important step toward sustainability.

This book has four main sections and a short Appendix. Section I reviews the problems with the ways conventional economics treats the natural environment and describes the alternative insights of ecological economics. Section II defines natural capital and describes its importance and what is happening to it. Section III introduces some concepts and policy tools relevant to the improved management of natural capital for a sustainable future. The Afterword briefly ponders some of the value questions raised by the preceding analysis of our ecological dilemma. And finally, the Appendix lists a few tools for those interested in taking personal and community action on the ideas and proposals described in the preceding pages.

Acknowledgments

We would like to thank the following individuals for various important contributions to this book: David Adamski, Ian Baldwin, Curtis Bohlen, Gundel Bowen, Ed DeBellevue, Anne Hambleton, Erika Harms, Robb Harrington, Dan Janzen, Patrick Kangas, Dennis King, Don Lotter, Peter May, Donella Meadows, Joni Praded, Jim Schley, Paul Temples, Frank Turaj, Gerrit van der Wees, Mathis Wackernagel, and Ben Watson. We especially thank Sandra Koskoff of ISEE for her diligence and patience during writing and production, Adrian Reilly for illustrations, and Fran Younger for graphics contributions.

The Authors

Dr. Robert Costanza is director of the University of Maryland Institute for Ecological Economics, and a professor in the Center for Environmental Studies, at Solomons, and in the Zoology Department at College Park. He received his Ph.D. in systems ecology (with a minor in economics) from the University of Florida in 1979. He also has a Masters degree in Architecture and Urban and Regional Planning from the University of Florida. Before coming to Maryland in 1988, he was on the faculty of the Coastal Ecology Institute and the Department of Marine Sciences at Louisiana State University in Baton Rouge, Louisiana. Dr. Costanza is co-founder of the International Society for Ecological Economics (ISEE) and chief editor of the society's journal, *Ecological Economics*.

John Cumberland is Professor Emeritus at the University of Maryland, where he served as Professor of Economics and Director of the Bureau of Business and Economic Research. His teaching, research, and publications have been primarily in the fields of environmental and natural resource economics. He is currently Senior Fellow at the University of Maryland Institute for Ecological Economics (IEE).

Herman Daly is the author of many works on ecological economics including *Steady State Economics* (1974). The most recent amplification of his ideas is found in the book he co-wrote with John Cobb, *For the Common Good* (1989). He is Associate Director for the University of Maryland Institute for Ecological Economics (IEE) and a Senior Research Professor in the School of Public Affairs at College Park. He is co-founder of the International Society for Ecological Economics and Associate Editor of *Ecological Economics*. He won both the Netherlands Royal Academy award and the Alternative Nobel Prize in 1996 for pioneering the new discipline of ecological economics.

Robert Goodland is the Environmental Advisor to the World Bank in Washington, D.C., and has published 17 books, mainly on tropical ecology, including *Race to Save the Tropics* (1990) and *Population, Technology, and Lifestyle: The Transition to Sustainability* with Herman Daly and S. El Serafy (1992). He was elected chair of the Ecological Society of America (Metropolitan) and the President of the International Association of Impact Assessment.

Richard Norgaard earned a Ph.D. in economics from the University of Chicago before going on to investigate the environmental problems of petroleum development in Alaska, hydroelectric dams in California, pesticide use in modern agriculture, and deforestation in the Amazon. He has been a member of the faculty of the University of California at Berkeley since 1970, where he is currently a Professor of Energy and Resources. He is President of the International Society for Ecological Economics (ISEE).

Thomas Prugh is an energy and environment writer and a senior energy analyst with a Washington, D.C.–area technology services company. He is the lead author of a forthcoming book on the study of the politics of sustainability.

THE ECOLOGICAL ECONOMICS PERSPECTIVE AND WHY IT'S NEEDED

1

Chapter 1

The Origins of Our Economic Worldview

The outdoors is what you have to pass through to get from
your apartment into a taxicab.
— Fran Lebowitz

In biological terms, humanity has succeeded to a fault. We take virtually
the entire Earth as our habitat, a claim no other species can make. By
learning to use nature ingeniously, we have invented ways of living that
allow us to imagine ourselves free of it.

However, as we carry on expanding our population and our artifacts,
we have begun to threaten the ecological integrity of that habitat. Humans
have had the power to disrupt local ecologies for tens of thousands of
years and have exercised it many times, through irrigation, overgrazing,
overhunting, forest cutting, river damming, city building, and so on. But
now the global scale, power, and penetration of human activity have
transformed our relationship to the Earth. Like the compulsive hand-

washer whose skin becomes raw and infected, we are finding that our labors to satisfy our needs and wants have begun to boomerang.

Among the roots of this problem is our way of thinking about ourselves, the world, and our place in it. Many of the troublesome assumptions underlying current values and attitudes toward the environment are embodied in conventional economic theory, which is the handmaiden of the industrial culture that has made high standards of living a reality for a few people and an ardent fantasy for billions more. Conventional (neoclassical) economics in turn rests on a very old view of the natural world. Early signs of it can be seen in the development of Western technology and the willingness to use it aggressively to manipulate nature (White, 1967).

Perhaps the classic example of this is the widespread use of the moldboard plow (see Figure 1.1) in northern Europe by about the 7th century. Where earlier plows only scratched the soil, requiring farmers to cross-plow their plots to loosen the soil, the moldboard plow employed a more powerful action that cut deep furrows and turned over the soil in one pass. The new plow was considerably more efficient, but its more important impact was a revolutionary change in social organization and accompanying views of the natural world.

The scratch plow (see Figure 1.2) could be pulled by two oxen, which many peasant farmers could afford individually. Families farmed as units and land was distributed in allotments capable of supporting a single family. However, the drag of the moldboard plow was so great that eight oxen were needed to pull it. Subsistence farmers, unable to afford so many oxen, pooled their teams and received land according to their contribution. Land distribution patterns thus came to depend less on the needs of families than on the imperatives of a technology. People's relation to the soil was also transformed as families' ties to particular plots of land were subordinated to their role in supporting the farming technology they used. Subtly, people were no longer tied to nature, but to a means of its exploitation.

Where did the exploitative tendency itself come from? Perhaps it was inevitable, once the eons-old hunter-gatherer way of life, a victim of its own success, gradually surrendered to agriculturalism. Thomas Hobbes may have characterized life in a state of nature as "solitary, poor, nasty, brutish and short," but in fact, hunter-gatherers typically made a relatively good living off the land at the cost of surprisingly little labor.* In contrast, the struggle to wrest a living from the natural world by farming appears

* Hunting and gathering worked so well that rising population pressures eventually required development of a system capable of greater output. See Gowdy, 1998 and Ponting, 1991.

Figure 1.1 Moldboard plow.

Figure 1.2 Scratch plow.

relatively precarious. For people living in societies that had made the
transition to settled agriculturalism, the surrounding unsettled lands must
have no longer seemed welcoming, as they would have to hunter-gatherers
who knew their secrets. Untamed wilderness would have been something
to be feared:

> If paradise was early man's greatest good, wilderness, as its
> antipode, was his greatest evil. In one condition the environ-
> ment, garden-like, ministered to his every desire. In the other
> it was at best indifferent, frequently dangerous, and always
> beyond control. And in fact it was with this latter condition that
> primitive man had to contend. At a time when there was no
> alternative, existence in the wilderness was forbidding indeed.
> Safety, happiness, and progress all seemed dependent on rising
> out of a wilderness situation. It became essential to gain control
> over nature (Nash, 1982, p. 9).*

   The Judeo-Christian tradition, particularly as it was expressed in Western
medieval Christianity, contributed to this exploitative tendency. "Be fruitful
and multiply, and fill the earth and subdue it," reads God's command to
humanity in Genesis, Chapter 1. The other creation story in Genesis, told
in Chapter 2, expresses the same idea: "Then the Lord God said, 'It is not
good that the man should be alone; I will make him a helper fit for him.'
So out of the ground the Lord God formed every beast of the field and
every bird of the air...." The idea of humanity's right to control creation
appears repeatedly in the Old Testament. It harmonized with the belief
that humans were made in God's image and that, alone among all the
creatures of the Earth, they largely shared in God's transcendence of nature.
Thus, although the rest of God's creation was also seen as having intrinsic
value, the stronger idea was that nature was meant for humanity's use.

---

* Now that we have mostly succeeded, we can indulge a few regrets: "One of
civilization's supreme ironies concerns the elimination of challenges, including fear,
hardship, and pain, that merely surviving in the precivilized world entailed. For
thousands of years after our race opted for a civilized existence, we dreamed of and
labored toward an escape from the anxieties of a wilderness condition only to find,
when we reached the promised land of supermarkets and air conditioners, that we
had forfeited something of great value" (Nash, 1982, p. 267). Wes Jackson notes the
irony that "wilderness is becoming an artifact of civilization. Civilization is all that
can save wilderness now" (Jackson, 1991).

Christianity also embodied a belief in endless progress that was not characteristic of Greek, Roman, or ancient Oriental civilizations. Tinkering with nature was thus encouraged by the notions that progress was natural and that humans had a divine license to make things better. The way was eased by Christianity's sharply different ideological attitude, compared with the animism and paganism it displaced, toward the manipulation of the natural world:

> If man believes that natural objects like stones, wind, water, and crops are moved by essentially arbitrary wills, either he will despair of manipulating nature to his own advantage or he will attempt to do this in the same way that he would attempt to manipulate his fellow man — that is, by attempts at verbal or symbolic communication, in the form of incantation and ritual. It is not until animism is replaced by an attitude which regards will as essentially and solely a property of the minds and souls of men, rather than of inanimate objects, that a scientific and technological attitude toward the world becomes possible. It is no accident therefore that the scientific transition originated in Western Europe, where the prevailing religion was an ethical monotheism, which either tended to concentrate the whole animistic enterprise in a single sacramental act of the Mass, as in Catholic Christianity, or which denied even this apparent remnant of animism by stressing that the operation of the will of God takes place principally in the souls of men, as in Protestant Christianity (Boulding, 1964, pp. 15,16).

By displacing or co-opting paganism, which had invested every tree, brook, hill, and creature with its own guardian spirit, Christianity sanctioned progress without concern for the feelings of natural objects (White, 1967; Daly and Cobb, 1989). In effect, nature no longer became spirit, but natural resources.

Natural resources eventually became the domain of economics, under the heading of land. As one of the three traditional factors of production (with labor and capital), land forms part of the economic Trinity. Since the advent of industrialization, however, land has diminished in importance compared to labor and capital, in terms of the attention given it in economic theory. (The importance of land as a factor of production in agricultural societies, on the other hand, is obvious.)

Land drifted out of the limelight gradually. In the 17th century, European economic thinkers began arguing whether land was an active or passive agent in the production of wealth. Later, the influential theories of David Ricardo (1772–1823) built on those of philosopher John Locke in asserting that land is passive and unimportant and makes no contribution to value. In this view, value (even of crops) derives only from expended labor. About the same time, economic thought in America, following Alexander Hamilton, began to hold that land was a form of capital. This meant that it did not need to be analyzed separately from capital in general. It was only a few short steps to the view of modern economics, which regards land (nature and natural resources) as "a mixture of space and expendable, or easily substitutable, capital. Both are treated as commodities ... subject to exchange in the marketplace and ... having their value determined exclusively in this exchange. Land is no longer a factor of production in any important sense" (Daly and Cobb, 1989, p. 111).

The assumption that different forms of capital are substitutable is absolutely crucial to this conclusion. It implies that production in general, whether of food, furniture, microchips, automobiles, perfume, steel, books, or any other item, need not depend on any minimum quantity or quality of natural resources. Conventional economics assumes that even drastic declines in natural resources can be compensated for by increasing other factors (labor and/or capital) within certain cost-imposed limits, and that doing so allows production to remain constant or even increase. This view thus allows a rather extreme measure of disregard for the integrity of whole ecological systems that enfold natural resources.

Classical Economics, Key Figures, Assumptions*

> One principle that is an ecological upsetter
> Is that if anything is good then more of it is better,
> And this misunderstanding sets us very, very wrong,
> For no relation in the world is linear for long.
> — Kenneth Boulding

The word economics, used in its roughly modern sense, first appeared in 1792. (In its sense of the science or art of household management it is at

---

* This section is based extensively on Costanza et al., 1997, *An Introduction to Ecological Economics*, Section II, and on other sources as noted.

least 400 years older.) When the discipline of economics evolved from moral philosphy in the 18th century, it was deeply concerned with how individual pursuits affected larger social goals. A key issue in particular was (and still is) whether personal greed and its satisfaction can work in the interests of society as a whole. Industrialization threw this issue into sharp relief, as the wholesale harnessing of fossil fuels vastly multiplied humanity's ability to convert natural resources into "wealth." Industrialization created the hope that the union of science and technology would lead to the mastery of nature and thus an age of plenty.

These wrenching changes did indeed create unprecedented opportunities for material advancement, at least for some. Shifting views about the morality of materialism helped justify the transformation. Before the Renaissance, material security was regarded as the reward for moral conduct. Afterward, it was seen as the prerequisite for it, i.e., moral progress required material progress. Scarcity caused greed and war, it was thought, and in general made it more difficult to lead a moral life.

Today, material progress has become a moral end in itself, the question of its relation to public and individual morality generally forgotten. Scientific progress is seen as both relentless and as the positive engine of change. The belief in the inevitability of breakthroughs and their ability to solve problems is deeply held.

This is the context in which economics, originally called political economy, evolved as a discipline. Several of its key early figures introduced ideas that remain central to our ecological dilemma.

The founder of modern economics, **Adam Smith** (1723–1790), was a moral philosopher whose first book (*The Theory of Moral Sentiments*) was on ethics (see Figure 1.3). However, Smith's reputation rests on his famous *Inquiry into the Nature and Causes of the Wealth of Nations*, published in 1776. In it, Smith argued that society is merely the sum of its individuals, that the social good is the sum of individual wants, and that markets automatically guide individual behavior to the common good (the "invisible hand"). One of the foundations of his theories — and the key to increasing production — was the division of labor, both among individuals and among nations (the doctrine of specialization). Another was that all the factors of production are freely mobile, i.e., that "production units and processes would remain small and nonspecific; ... that small producers would meet small consumers in the marketplace, that they would have equivalent power and information, and that no nuisance effects — externalities — would spill over onto innocent bystanders" (Henderson, 1981, pp. 191-2).

Smith's model was atomistic and mechanistic, a viewpoint that economics has retained into the neoclassical era (see "The birth of neoclas

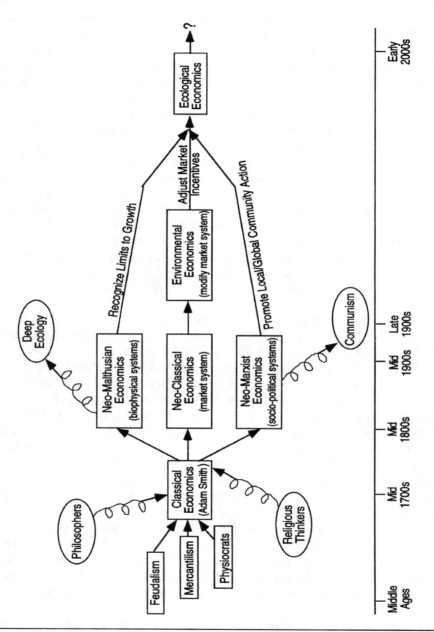

Figure 1.3   Ecological economics; a family tree. Source: Dennis King (1992). After Heilbroner (1980) prior to Adam Smith and Colby (1990) afterwards.

sicism," subsequently). In contrast, the ecological view is that people's identities derive from their relations to others, as do their ideas of what is needed and important for a meaningful life, and that communities' histories are important. Ironically, Smith's roots in moral philosophy can be seen in his concern about the social problems caused by the division of labor, including isolation, the disruption of families, and the psychological damage inflicted by doing the same thing over and over, which Smith believed could affect the capacity for moral judgment (Skinner, 1988).

Nevertheless, Smith's economics made morality less important by creating an economic rationale for regarding individuals as freely mobile agents rather than individuals living as members of a community, linked by its moral and social bonds. Smith's theories helped make it possible to argue that the social good could be promoted even when people do only what suits them best as individuals.

Thomas R. Malthus (1766–1834) was a gentle parson and mathematician made famous by his grim *Essay on the Principle of Population*. To sum up this essay in his own words, "[p]opulation, when unchecked, increases in a geometrical ratio. Subsistence increases only in an arithmetic ratio.... By that law of our nature which makes food necessary to the life of man, the effects of these two unequal powers must be kept equal. This implies a strong and constantly operating check on population from the difficulty of subsistence" (cited in Black, 1988, p. 224).*

Malthus also believed that wages would settle at subsistence levels. Temporarily rising wages (i.e., improved well-being) would lead to a surplus of workers, which would then cause wages to plunge below subsistence levels, eliminating the surplus through die-off and restoring equilibrium in the supply and demand for labor (Henderson, 1981). Malthus developed his law of diminishing returns in connection with land-use practices, anticipating the ecological view that increased inputs to agricultural land will reach a point at which yields do not keep up and could even decline.

Malthus' model of population growth is too simple to be true in general, although it has been true in certain times and places. It may have helped create indifference to the problems of population, by creating a straw-man specter of doom in which human population would increase exponentially, outstripping the merely arithmetic growth of food supplies, and then crash because of food wars and starvation. This caricature made it possible to dismiss the more complex problems caused by human population growth. Malthus' model may even have invited the arguments of the so-called

---

* Malthus did not claim that productivity increases were impossible, only that population would tend to outpace them. See Ayres (1993).

cornucopians and technological optimists that "the more, the merrier;" that more people will solve the problems that an excess of people have caused. Ecological economics argues that this position is counterintuitive at best and carries a heavy burden of proof.

We have already been introduced to **David Ricardo**. Besides his ideas on the economic value of land as a factor of production, Ricardo is also important (for our purposes) for his theory of rent. Briefly, Ricardo said that the best land, which produces the most food for the least work, is farmed first. As population increases, people must cultivate additional, less fertile land, which requires more labor per unit of food produced. Food prices therefore rise to cover the cost of the extra labor. The best land then earns a return over and above its cost of production, i.e., a rent. At the same time, the higher food prices create an incentive to work the good land more intensively.

Ricardo's model helps explain why increasing population forces the conversion of natural landscapes to farmland, and why modern agriculture has turned so enthusiastically to the intensive use of agricultural chemicals and heavy machinery dependent on fossil fuels. What Ricardo could not have anticipated is the consequences of agricultural intensification. There is no such thing as a free lunch, and getting more food out of a given plot of land means boosting inputs and working the land harder. This approach depletes existing stocks of nonrenewable resources (petroleum and natural gas, which are used as fuels and feedstocks for agricultural chemicals), accelerates erosion, and, through use of monocultures, weakens the gene pool.

John Stuart Mill (1806-1873) further explored Adam Smith's ideas about the connections between individual behavior and social well-being. In doing so, Mill developed one of modern political conservatism's most important theses: that competitive markets are critical for preserving individual liberty. Interestingly, Mill did not flinch from extending his ideas about liberty to women. He argued for equality of the sexes within the family as well as women's "admissibility to all the functions and occupations hitherto retained as the monopoly of the stronger sex," and said that "their disabilities elsewhere are only clung to in order to maintain their subordination in domestic life" (*The Subjection of Women*, cited in Blau, 1991, p. 291). His views anticipate current concerns about the issues of women's freedom and status and their effect on environmental degradation, especially in the developing world.

Today, conventional economics generally rejects one of Mill's critical insights, namely, that *perpetual growth in material well-being was neither possible nor desirable.* Mill believed that economies naturally mature into a steady state, leaving people free to pursue nonmaterialistic goals. In fact, the classical economists in general — Smith, Malthus, Ricardo, Mill — all regarded long-run growth as unlikely. In Mill's vision, this was clearly not all bad, but the others were more pessimistic. Smith, Malthus, and Ricardo thought various limits would eventually retard and stop growth, with the process ending at subsistence-level living for most people (Pearce and Turner, 1990).

Mill's view of the steady-state economy resonates profoundly with the lessons of biology, which teaches that change is unceasing but growth is not. Moreover, Mill not only saw the steady state as natural, he foresaw and understood its implications for the social distribution of wealth. He knew that distribution — how the fruits of economic activity are divided — was a political process, not an economic one: "The things once there, mankind, individually or collectively, can place them at the disposal of whomsoever they please and whatever terms. Even what a person has produced by his individual toil, unaided by anyone, he cannot keep unless by permission of society" (*Principles of Political Economy*, cited in Henderson, 1981, p. 205). When the stationary state was reached and no more accumulation was possible, distribution would become the only issue.

The Birth of Neoclassicism

Modern economics may have ignored Mill's doubts about perpetual growth, but it owes him a great debt for its current theoretical structure:

> In 1848, Mill published his *Principles of Political Economy*, a Herculean reassessment [of economics] that came to a radical conclusion: economics had only one province: production and the scarcity of natural means. This narrowed the focus of political economy to a "pure economics," later called "neoclassical," which allowed a more detailed focus on the economic core process while excluding social (not to mention environmental) variables in an analogue of the controlled experiments of the physical sciences. After Mill, economics became split between the neoclassical, mathematical, "scientific" approach and the more policy-oriented "art" of broader social speculation (Henderson, 1981, p. 204).

The field of political economy thus divided into two broad camps around the middle of the 1800s. One focused on "broader social speculation" and concerned itself with social structures, value systems, and relationships among the classes. This group included Karl Marx and other reformers, including socialists, anarchists, and utopians, as well as a handful of classical economists such as Mill. It became increasingly marginal to the main thrust of the narrowing discipline of economics, whose banner was carried by the emerging neoclassical school.

Neoclassical economics, which can be dated to approximately 1870, ushered in the modern economic worldview. Neoclassical economics differed in important ways from the classical school that preceded it. First, the value of a good was no longer seen as dependent on the labor expended to produce it, but instead derived from its scarcity. The good's price (value) was seen as determined by the interaction of supply and demand. Neoclassical economics also deemphasized concerns about long-term growth and stressed marginal analysis, the study of the relationships between small changes in prices and quantities as economic activity is extended by small amounts (Pearce and Turner, 1990).

Perhaps most importantly, while classical economics was closely allied to moral philosophy, neoclassical economics was meant to be value-free. Casting itself as a scientific enterprise, neoclassical economics tried to develop laws to describe economic activity. People were rather narrowly defined as rational and egotistic agents who merely try to satisfy their wants, and, in so doing, incidentally improve collective social welfare. They were not seen as having an ethical or moral dimension. The neoclassical world works like this:

> [W]henever any individual is not satisfied with some aspect of his current situation (say, not consuming enough bread), he ... enters the market and competes with other buyers by bidding up the price ..., thereby creating an incentive for at least one producer to sell to him rather than anyone else. This process of increasing the going market price raises the average selling price and thereby indicates to producers that more is wanted, and to other buyers that they should consider cheaper substitutes ... . If sufficient time is allowed for all such market activity to be worked out, eventually all individuals will be satisfied relative to what they can afford (to afford any more may mean they would have to work more than they considered optimal) (Boland, 1988, p. 288).

In theory, this sort of competitive market activity continues over the long term until a level of maximum efficiency is reached, called the *Pareto optimum* (named for Vilfredo Pareto, a 19th century Italian mathematical economist). This is the point at which nothing else can be done to make anyone better off without making someone else worse off. In this tidy model it is assumed that all players have perfect information, are free of other constraints, and are not motivated by anything other than maximizing personal "utility" (satisfaction of wants and needs). Mostly absent are institutions — other than the organic institution of the market — and the roles they play in socializing and moderating human behavior.

Economics thus strove to be a science. Fatefully, it chose to model itself upon Newtonian physics. The reigning queen of the natural sciences, physics was the littermate of industrialization and had made the very heavens themselves understandable. Its predictive power seemed unparalleled. There was, however, another road (not taken): evolutionary biology. Charles Darwin had published *Origin of Species* in 1859, and the theory of evolution by natural selection was to prove equally powerful, in its own way, as the laws of celestial mechanics.

These models represented profoundly different outlooks. By following physics, economics chose to emphasize the search for lawful relationships describing current economic behavior rather than those that might describe the ways economic systems change. The classical economists, as we saw earlier, believed that growth was fated to stop. They knew that, however valuable were the models used to describe the economic system during its growth phase, those models might be misleading in the post-growth phase. The classical economists understood that the behavior of economic systems was *contingent*: it depended on history and circumstances. By emulating the model of Newtonian physics, however, neoclassical economists ignored this contingency and adopted a deterministic outlook. Leon Walras, one of the founders of neoclassical economics, talked of people as "economic molecules" (Henderson, 1981) and consciously set out to place economics on a scientific plane as lofty as Newton's achievement with celestial mechanics (Daly and Cobb, 1989).

A key result was the search for mathematical rigor, an enterprise at which economics has excelled. A second result was a tendency to abstract from the real world. Nobel economist Wassily Leontief once surveyed the articles appearing in nine years' issues of *American Economic Review*, a major economics journal, and found that only about 1% reported on studies for which the authors had helped to gather the data and fully half were discussions of mathematical models that relied on no data at all (Leontief, 1982). "Nothing reveals the aversion of the great majority of the present-day academic economists for systematic empirical inquiry more than the methodological devices that they employ to avoid or cut short the use of concrete factual information," Leontief wrote in this penetrating critique of academic economics. In his book *Complexity*, Mitchell Waldrop describes a Sante Fe Institute meeting of some of the nation's top physicists and economists, at which the latter spoke on standard neoclassical theory.

"... [A]s the theorems and proofs marched across the overhead projection screen, the physicists could only be awestruck at their counterparts' mathematical prowess — awestruck and appalled. ...'They were almost too good,' says one young physicist.... 'It seemed as though they were dazzling themselves with fancy mathematics, until they really couldn't see the forest for the trees'" (1992, p. 140).

Driven to Abstraction: The Neoclassical Legacy

The historical development of economic thought thus led to the dominance of a view, neoclassical economics, that emphasizes theory over observation and ignores data that cannot be easily reconciled to theory. Opting for purity over concreteness, the mainstream view (despite challenges from resource and environmental economists) has clung to a number of questionable assumptions. The key ones are summarized here.

- The environment (a.k.a. natural resources, "land," natural capital) is seen as part of the human economy — a factor of production — rather than the other way around. The role of the environment in supplying materials, services, and waste sinks, which together make the human economy possible, is largely ignored.

- Even when they are treated as valuable, the materials and services provided by the environment are not seen as indispensable. It is assumed that human-made capital can be substituted for natural resources virtually across the board. Nobel economist Robert Solow once claimed that "...the world can, in effect, get along without natural resources" (cited in Rees and Wackernagel, 1994).*

- This doctrine of infinite substitutability leads naturally to the rejection of any limits on growth, even thermodynamic ones. Human technology is seen as perpetually able to overcome scarcity through substitution of human-made resources for natural resources and through ever-higher efficiencies. Ceaseless growth is seen as not only possible but as the best, perhaps the only, solution to poverty and environmental degradation. In *The Economist*'s 150th anniversary issue, which speculated about the world economy over the

---

* Solow has since moderated his views on this subject, but many other mainstream economists have not.

next 150 years, conservative economist C. Fred Bergsten predicted that "[s]tandards of living will rise sharply almost everywhere, even as the global population rises to between 12 billion and 15 billion, as technology continues to expand exponentially and virtually all regions adopt the policy reforms that began to proliferate in the late 20th century" (Bergsten, 1993, p. 61). Continued economic growth and population expansion are taken as merely the logical extrapolation of past trends.

- The welfare of human society is best served by the view of people as "human molecules" who, by pursuing their own interests through the market, inevitably promote the general good. There is little need to consider things from the ecological point of view and embrace the notions of interaction, interdependence, community, and the noneconomic relations people enjoy with each other and with the natural world.

While these assumptions have worked reasonably well for many people, the circumstances in which they arose have disappeared. Neoclassical economic thought developed near the end of a very long period when the world was essentially empty of human beings, their artifacts, and their waste. Now the world is, if not actually spilling over, perilously close to being full. The continued soundness of neoclassical assumptions for our times is doubtful. The next chapter discusses another way of looking at things.

Chapter 2

The Ecological Economics Perspective*

Economics can be seen as the ecology of man; ecology as
the study of the economy of nature.
— Marston Bates, *The Forest and the Sea*

The fundamental error of the dominant economic worldview is to treat
land (the environment) as merely a factor of production (and one of
declining importance, at that). In effect, this outlook locates the environment within, and subordinates it to, the human economy (see Figure 2.1).

From the ecological economics point of view, this is backward. We
have been looking at the environment and its resources through the wrong
end of the telescope. By turning the telescope around and looking through

---

* This chapter is based extensively on Costanza et al., 1997, *An Introduction to
Ecological Economics*, Section III, and on other sources as noted.

19

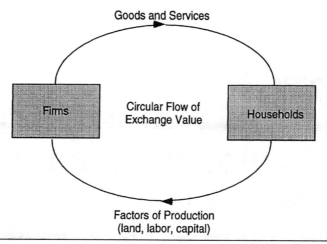

Figure 2.1   The neoclassical model of the economy.

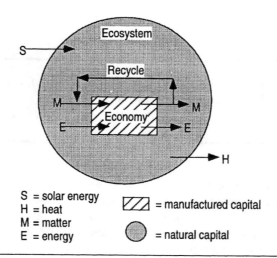

Figure 2.2   The economy as an open subsystem of the global ecosystem.

the proper end, the importance of the natural world is suddenly and rightly magnified. We can see that in fact the natural environment encompasses the economy (see Figure 2.2). The economy is an open system within the ecosphere, importing useful resources from the natural world and exporting wastes back to it (see "Entropy and Economics," pg. 40). Human beings appropriate the resources provided by the ecosphere (oil, timber, minerals, fish, crops, etc.) and turn them into food or artifacts (cars, Cuisinarts, compact discs) using labor and manufactured capital (machine tools, factories, transportation systems, etc.). This produces wastes (pollution) that the environment absorbs (up to a point). The environment also provides services (oxygen generation, water purification, soil creation, and so on; see Section II) that we must have but cannot supply. All economic activity is thus enabled by the natural environment. All economic production is actually consumption—of natural resources (Rees and Wackernagel, 1994).

Once we admit that the human economy nests within the ecosphere and cannot exist outside it, several concepts neglected or misunderstood by neoclassical economics become critically important. These ideas, beginning with the coevolutionary paradigm, form the core of the ecological economics perspective.

The Coevolutionary Paradigm*

In offering an alternative to the conceptual schemes of neoclassical economics, ecological economics emphasizes the coevolutionary development of human beings and the natural world. It is important to understand how thoroughly embedded we are in the natural world, how much we are a product of it as well as vice versa. Despite the occasional hurricane, earthquake, or drought that punctures the veneer of civilization, we generally imagine that we have achieved control of nature. But it is much more accurate to say that humans and their systems (knowledge, artifacts, cultures, and technologies) *symmetrically coevolve* with natural systems. These systems have developed in relation to one another over very long periods of time in a dance of mutual interaction, influence, change, and selection. Innovation, discovery, and random change occur continuously in each system. These in turn increase or decrease pressures on various pieces of the other systems. Thus the state or identity of each system at any given time reflects the historical influence of the others.

This has been going on for eons. Consider biodiversity and its loss, for example. Current rates of species loss due to human economic activity are shockingly high (see Chapter 4). But our hunting, gathering, and farming ancestors shared the same impulse to shape the world around them, even if they lacked the powerful technologies at our disposal today. They were equally swept up in the dance of coevolution:

---

* This section is based largely on Costanza et al.,1997 and Norgaard, 1994.

Hunting and gathering ... put selective pressure on the species hunted and gathered. People in hunting and gathering societies transformed habitats to favor particular game and plant products. At the same time, hunting and gathering selected for effective hunters and gatherers who were more likely to survive and to support a larger number of offspring.

...Agriculture has been a more deliberate and intensive effort with far more significant impacts.... Planting and watering, hand weeding, plowing, flooding, and burning are direct means of favoring productive species, reducing the competition for nutrients by "weeds," and nurturing species that complement each other. ...Later the selective pressure on the biological system was heavily influenced by the agricultural practices of deforesting and clearing land, weeding out some species and encouraging others, enriching soil, and eventually irrigating. These environmental transformations facilitated the evolution of species unable to survive without people. Agriculture, in turn, selected for different perceptive abilities and physical strengths among humans than had hunting and gathering (Norgaard, 1994, p. 89).

These examples show how the coevolution of nature and humans means the coevolution of nature and culture as well. Because humans are a social species, our survival depends on the success of the groups to which we belong, and a group's survival is tightly bound up with its culture. Culture helps sustain social organization and give meaning and shape to life. Groups confronted with rapid change in the natural world may suffer major cultural trauma; the tragedy of the plains Indians in the face of European encroachment and destruction of the once-vast buffalo herds comes to mind. And as we saw in Chapter 1, cultural beliefs can powerfully alter the face of the natural world.

Modernity has not abolished coevolution. For example, since World War II farmers have made greater and greater use of chemical pesticides to control crop-robbing insects. In turn, the target insect populations have bred resistance, leading to cycles of initial decline in pest populations followed by the rebound of tougher bugs. This has selected for (forced the development of) new and more lethal pesticides. In the meantime, evidence has mounted about rising concentrations of pesticides in groundwater and other problems. The bugs keep developing new strains. Eventually, the process may select for new ways to think about controlling pests, such as integrated pest management. This would be a step toward an important cultural shift in values, if the impulse to swat the bugs with powerful chemicals begins to give way to a more harmonious approach involving more subtle and sustainable techniques, such as intercropping and encouragement of pest-insect predators. This process is mediated by the roles and political interests of the various human actors, e.g., the farmers, pesticide manufacturers, politicians, environmentalists, consumers, and so on. In short, values, knowledge, human organizations, the environment, and technology coevolve with one another.

The development of chemical pesticides shows that natural processes can be bypassed or overridden for a time, allowing human culture to go off in a developmental direction of its own, apparently liberated from the coevolutionary demands and constraints of natural systems. Perhaps the most important example of this is our discovery of how to use fossil fuels to intensify our agriculture and transform our extraction and use of resources in general. But this freedom is illusory: "…[D]evelopment based on fossil hydrocarbons allowed individuals to control their immediate environments for the short run while shifting environmental impacts, in ways that have proven difficult to comprehend, to broader and broader publics, indeed the global polity, and on to future generations" (Costanza et al., 1997).

There are at least two potent implications of the coevolutionary paradigm for the prospects of sustainability. First, evolution does not necessarily mean progress. We are seduced by the idea that humankind is the final end of evolution and that everything will therefore work out. But the process has not stopped; coevolution implies the continuous and endless mutual adaptation of systems to changes in related systems. Nor are we, by many measures, the most successful species. Insects, for example, vastly outnumber and outweigh human beings and are arguably at least as tough, adaptable, and resilient as we. They are also far older. On the other hand, mere longevity and success is no guarantee of perpetual survival; witness the dinosaurs. Whatever we are, humans are not exempt from the coevolutionary processes that have exalted some species and deposed others.

Second, the coevolutionary nature of our relationships with the natural world means that there is an irreducible minimum of unpredictability about them. The natural world is not a machine and the complex systems we inhabit and interact with cannot be described mechanistically. Random and/or unforeseeable introductions of new elements constantly occur, not only in biological systems (through mutation and selection) but in cultural systems as well. Machine behavior is predictable because machines lack this creative dynamism. The parts of a machine and their relationships to each other remain static. Natural systems are different.

Our ability to predict and control nature is therefore *inherently* limited. It is not just temporarily limited by lack of data, scientists to gather it, and funding to support the scientists (though more of those things would not necessarily be bad), but permanently and intrinsically limited by the nature of nature. Thus, the coevolutionary paradigm strongly implies the critical importance of adaptability, the need to preserve multiple ways of relating to the natural world, and the wisdom of choosing policies that preserve future options in the face of uncertainty.

Scale, Throughput, and Carrying Capacity

How big is the economy relative to the global ecosystem that sustains it? How big can the economy become without destroying the ecosphere and thus itself? Finally, how big should it be? That is, what is the optimal, or best, size for the human economy? Is it conceivable that the largest possible economy may not be best in the long run?

These are the problems of scale (Daly, 1992). They are hardly addressed at all by neoclassical economics, which has instead focused heavily on the problems of allocation, i.e., how best to divide resources among various uses. (It turns out that markets and prices are very efficient at determining optimal allocations, far better than the bureaucrats of centrally planned economies.) But paying attention only to allocation and not scale is like overloading a rowboat; no matter how carefully and evenly the passengers are seated fore and aft, port and starboard, the boat will sink if their combined weights are too great for the boat's carrying capacity (Daly, 1991a).

The crucial feature of economic scale is throughput, which is the volume of materials flowing from the environment through the economy and back to the environment as wastes. Throughput depends primarily on the number of people living on the Earth and their per-capita resource use. Economic growth (higher output) requires greater throughput and thus incurs proportionally greater costs.

In some ways we can compare the process to an automobile engine. Running the engine at higher speeds or greater loads results in higher power output, but also requires greater input in the form of air and fuel. Running the engine too fast and hard can cause catastrophic damage. Even at sub-catastrophic levels, the harder the engine is run, the faster it wears out. The economy/ecosystem likewise can be "run" harder to produce more economic output, but to do so results in more rapid depletion of resource stocks, greater volumes of wastes expelled to the environment (perhaps at higher rates than they can be absorbed or made harmless), and greater "wear," in the form of eroded and salinized soils, expended groundwater, biodiversity loss, and so on.

This analogy breaks down if pushed too far. For one thing, an engine has no powers of self-renewal, while the ecosystem does (though they can be overwhelmed). Another important difference is the fact that economic growth (greater throughput) means that the economic system gets bigger with respect to its "fuel system," i.e., the ecosystem that supports it. If the old adage, "Buy land — they ain't making any more of it," is true for real estate, it is doubly true here: the ecosystem is utterly finite and cannot expand.* If the economy grows too large, it will consume its support system and both will collapse.

How big is the world economy compared to the global ecosystem? It is difficult to measure, but one approach is to calculate how much of the Earth's net primary productivity (NPP) is used or otherwise taken over by human beings (Vitousek et al., 1986). NPP is the output of primary producers, which include land-based green plants and aquatic microorganisms called phytoplankton. Primary producers are photosynthesizers: they use sunlight to live and make more plant mass. This mass supports nearly all other living things, including humans, since in the end we all live off it by using it for food and/or habitat, either directly or indirectly. Because the Earth is a finite ecosystem that receives a finite flow of solar energy, it is capable of only a certain maximum rate of net primary production. We share this bounty with all other species. To the extent that we use more, less is left for all other creatures.

Humans appropriate much of the Earth's potential NPP in a variety of ways. First, we consume it directly. People eat plants as well as domesticated animals that eat plants. We eat fish that also depend, one or two steps down in the food pyramid, on primary producers such as phytoplankton. We harvest trees for lumber, fiber, fuelwood, and so on.

Second, we do things that reduce the Earth's primary production potential. For example, we convert natural ecosystems to agricultural systems that tend to be less productive biologically (in part because most crops are annuals and most nondomesticated plants are perennials). We

---

* Utopians who envision the expansion of our habitat by colonization of the solar system and outer space as both human destiny and a solution to our earthly problems might pause and think about the costs. Even assuming that suitable worlds were accessible—and there is no evidence that any other place in the solar system could support significant numbers of people without heroic engineering intervention—it's well to remember that the human population of Earth is growing by about 85 million people annually. Even to keep global population constant thus means exporting roughly the population of Mexico every year. The costs would be, well, astronomical.

The Mostly Invisible City*

Mainstream economics generally either ignores the idea of carrying capacity or dismisses it as irrelevant by arguing that limits, if they exist, are remote and/or that human ingenuity can transcend them. But the importance of carrying capacity can be brought closer to home by looking at how cities survive.

However, large and sprawling, cities are mostly not where they seem to be. Urban areas are utterly dependent for survival on consuming the output and waste absorption capacities of vast areas that lie outside their nominal limits:

> ...[P]reliminary data for industrial cities suggest that *per-capita* primary consumption of food, wood products, fuel, and waste-processing capacity co-opts on a continuous basis several hectares [a hectare is about 2.5 acres] of productive ecosystem....
> ...[T]he land "consumed" by urban regions is typically at least an order of magnitude greater than that contained within the usual political boundaries or the associated built-up area. However brilliant its economic star, *every city is an entropic black hole*

drawing on the concentrated material resources and low-entropy production (see "Entropy and Economics," p. 40) of a vast and scattered hinterland many times the size of the city itself. ...In ecological terms, the city is a node of pure consumption existing parasitically on an extensive external resource base (Rees and Wackernagel, 1994, pp. 370, 379).

Consider the ecological "footprint" of the region around Vancouver, British Columbia (see Figure 2.3). Conservative calculations suggest that each of the 1.7 million residents consumes the food output of 1.1 hectares, the forest-products output of 0.5 hectares and the fossil-energy output (derived from the area needed to produce an equivalent amount of renewable fuel) of 3.5 hectares, for a total of 8.7 million hectares. The region itself is only 400,000 hectares. The region's ecological footprint is thus 22 times larger than the region itself.

Cities not only borrow carrying capacity from elsewhere, they also borrow it from the past (when burning fossil fuels, for example) and from the future (by drawing down natural capital stocks for current consumption and leaving less for the generations to come). Mainstream economics and the development policies based upon it take little notice of these facts, but they have important implications for sustainability in general and for international trade in particular (see Section III).

---

* This section is based on Rees and Wackernagel, 1994, in Jansson et al., 1994, *Investing in Natural Capital: The Ecological Economics Approach to Sustainability.*

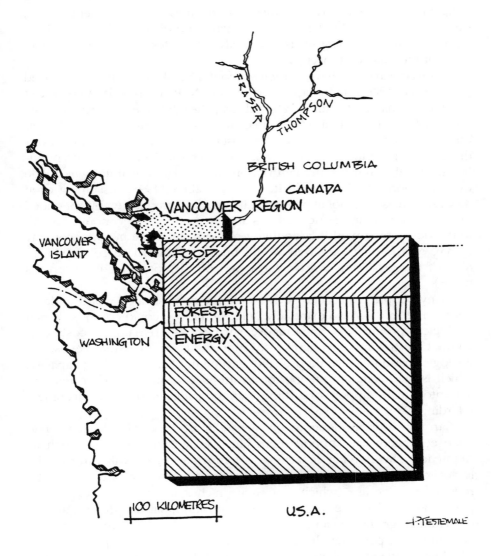

Figure 2.3   The ecological footprint of the Vancouver area is 22 times larger than the area itself.

convert forests to pastureland, which is also less productive. Farming marginal lands such as dry savannahs sometimes leads to desertification, further lowering total potential productivity. And of course building cities, highways, and shopping malls essentially obliterates the primary productivity of the land they occupy.

In these ways, we have cornered about 40% of the Earth's land-based NPP (about 25% of land-based and aquatic NPP combined) (Vitousek et al., 1986). That is, by the NPP yardstick, the global ecosystem devotes about 40% of its land-based resources to human beings. It is not clear at this time whether even this share is sustainable in the long term. Nevertheless, we are rapidly increasing it. According to United Nations population growth estimates, the Earth's human population will grow to a little over 8 billion by 2025 (World Resources Institute, 1998). Assuming even modest increases in global per-capita living standards (driven by continued overconsumption in the rich nations and the hunger for economic growth in the developing world), a doubling of the scale of the human economy in 35 years is plausible. That would bring the human share of NPP to 80% (if the process is strictly linear and no disruptive feedback effects occur).

Many people would find such a radically domesticated world undesirable to live in. It is also extremely unlikely that we could co-opt so much ecological productivity and impose such stress on natural ecosystems without destroying their ability to support us. Such extensive use of ecological resources would probably exceed the Earth's carrying capacity, which is the maximum number of creatures (humans and other animals) a territory can support over time without damage to the environment (Hardin, 1986).

Carrying capacity depends on many things. A rich, fertile, temperate zone ecosystem can support more life than a cold, arid place such as Antarctica. The carrying capacity of a temperate region will generally be lower in winter than in summer. The carrying capacity of an area can be reduced by overuse (which can create niches for weeds where edible plants once thrived), erode soils, increase runoff, silt up streams, and so on. Human ignorance can reduce an area's carrying capacity dramatically and, in effect, permanently. The valley of the Tigris and Euphrates rivers (the hypothesized inspiration for the myth of the Garden of Eden) remains a bleak place two thousand years after agricultural mismanagement ruined it (Hardin, 1986).

By ignoring carrying capacity, it is possible to make astonishing estimates of the number of people the Earth can support. Thirty years ago, British physicist J.H. Fremlin estimated that the total human population on Earth could be allowed to reach 60 quadrillion people, about 120 per square meter of the planet's surface area. It would take about 900 years with the population doubling every 37 years. Even then, in this view, the only obstacle to further growth would be the problem of getting rid of the excess heat generated by all those bodies and their vast life-support machinery.

Fremlin's vision is either satire or delusion. Everyone would be housed in a two-thousand story structure that covered the entire surface of the Earth; the bottom half would be living quarters and the top half, refrigeration and food-production machinery. Fremlin assumed the complete elimination of land-based wildlife; replacement of all ocean-dwelling wildlife with the most efficient photosynthesizing microorganisms (to increase food yields); use of large orbiting reflectors to light the poles and the dark side of the planet, thus increasing the harvest of sunlight and warming the whole Earth to equatorial temperatures; direct synthesis of food from recycled wastes and corpses; and, ultimately, roofing over the oceans and covering the entire surface of the planet with the mammoth building mentioned previously. The required raw materials would be taken from the oceans and the top 10 kilometers of the Earth's surface. No one would be able to travel more than a few hundred meters in any direction, but even that short tether would allow access to 10 million people, not counting communication by video-phone. Exercise, which would produce too much heat, would be largely outlawed.

Fremlin's vision may seem dystopian, but he offers this comfort: "The extrapolation from the present life of a car-owning, flat-dwelling office worker to such an existence might well be less than from that of the neolithic hunter to that of the aforesaid office-worker" (Fremlin, 1964, p. 287).

More plausible estimates of the Earth's carrying capacity for human populations have placed it at various levels from 7.5 billion to 50 billion. It is difficult to be more precise, partly because of our ignorance about the possibilities for stretching food supplies and how much we can co-opt natural ecosystems without sawing off the ecological limb we sit on. But as the Fremlin scenario suggests, the maximum supportable population is probably not the optimal population. The main reason is culture.

When applied to animal populations, carrying capacity can be discussed strictly in terms of the number of animals an area will support. An animal population generally grows until food supplies are exceeded, at which point the population declines, often sharply, through starvation and vulnerability to disease. It cannot recover until the food supply recovers. The population will tend to oscillate in this way around some equilibrium point.

However, animals differ from human beings in that they lack both class distinctions and culture. Unlike all other species, human beings are rarely content to live at the subsistence level. Once the problems of survival are solved, we turn quickly to inventing and acquiring centrally heated houses, disposable cameras, good Bordeaux, automobiles, and Club Med vacations. Most humans consume more resources than they need merely to survive, in order to satisfy wants. The rich consume vastly more than the poor. The impact of the human population on the environment does not depend only upon sheer numbers, but upon the per-capita level of affluence as well.

Thus the carrying capacity of the Earth for humans is only partly a biological issue; it is also a cultural and political one with psychological and spiritual elements. If everyone insists on living as affluent Americans, Swedes, and Swiss do, then the carrying capacity limit will be reached at much lower levels of population than if everyone lives as lightly on the Earth as does the average citizen of Burundi or Bangladesh. Since, in practical terms, living lightly often means living wretchedly, the developing world is determined to grow as rapidly as possible. And since the wealthy North (Europe, Japan, the United States, and Canada) has already appropriated a disproportionate per-capita share of the available resources, a similar scale of economic growth in the developing world will impose dangerous stresses on the global ecosystem. (The U.S., with about 5% of the world's population, uses about 25% of the world's energy resources. Concentrating all the rest would allow only another 15% of the global population to live the energy-lavish life of Americans.)

In fact, such growth can only take place by drawing down the stocks of the world's natural resources, which is much like drawing money out of a trust fund rather than living off the interest it generates. A large fund may yield enough interest for a decent living, *and will do so indefinitely, if the principal is left alone.* But as repeated withdrawals cause the fund to dwindle, the interest declines as well. Eventually the fund is empty and cannot produce any flow of interest.

Likewise with our stocks of natural resources. In our rush to maximize the scale of the global economy, we are eating our seed corn, consuming both the stocks and the flows they generate, rather than living sustainably off the flows. We are thus forced to confront grave geopolitical questions of overconsumption, population control, and the redistribution of existing wealth. And in addition to these compelling practical problems, there is another outcome of scale maximization that we would probably regret: the inevitable necessity of destroying or transforming every bit of untrammeled nature in the frantic drive to make it all "productive." John Stuart Mill spoke eloquently about this sad prospect 140 years ago:

Nor is there much satisfaction in contemplating the world with nothing left to the spontaneous activity of nature; with every rood of land brought into cultivation, which is capable of growing food for human beings; every flowery waste or natural pasture plowed up, all quadrupeds or birds which are not domesticated for man's use exterminated as his rivals for food, every hedgerow or superfluous tree rooted out, and scarcely a place left where a wild shrub or flower could grow without being eradicated as a weed in the name of improved agriculture. If the earth must lose that great portion of its pleasantness which it owes to things that the unlimited increase of wealth and population would extirpate from it, for the mere purpose of enabling it to support a larger, but not a happier or better population, I sincerely hope, for the sake of posterity, that they will be content to be stationary, long before necessity compels them to it (Mill, 1969, p. 750).

Substitutability vs. Complementarity

Describing natural resources in terms of stocks and flows invites some new terminology. It is useful to think of existing stocks of natural resources (traditionally called "land") as natural capital, since it functions in the natural world in much the same way as manufactured or reproducible capital does in the human economy. Just as manufactured capital in the form of factories and machine tools produces a flow of toasters and bicycles, natural capital in the form of forests and oceans produces a flow of timber and fish (see Figure 2.4).

It is obvious that, to some extent, the three traditional factors of production (land, labor, and capital, meaning natural resources, labor, and manufactured capital) can be substituted for one another. The modern economy offers many examples that illustrate both the principle and the sometimes-pointed consequences of such substitution, from the elimination of jobs through factory automation (replacing labor with capital) to substituting tractors and combine harvesters for farm workers (ditto) and boosting agricultural output through high-chemical-input, intensive farming of good land while allowing marginal land to lie idle (replacing land with capital).

As noted in Chapter 1, the substitutability of manufactured capital for natural capital is one of the most momentous assumptions of neoclassical economics. Here is how mainstream economists William Nordhaus and James Tobin (a Nobelist) put it:

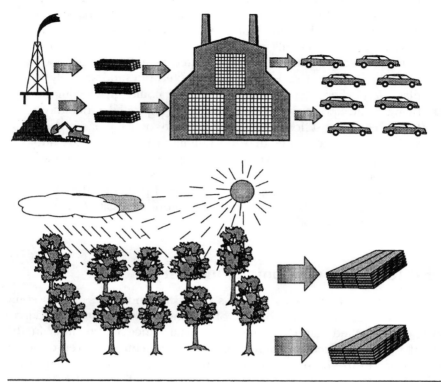

Figure 2.4   Stocks and flows.

The prevailing standard model of growth assumes that there
are no limits on the feasibility of expanding the supplies of
non-human agents of production. It is basically a two-factor
model in which production depends only on labor and reproducible [manufactured] capital. Land and resources, the third
member of the classical triad, have generally been dropped.
...The tacit justification has been that reproducible capital is a
near perfect substitute for land and other exhaustible resources
(cited in Costanza and Daly, 1992).

To assume that labor and manufactured capital can be substituted *ad
infinitum* for natural capital leads inescapably to the conclusion that the
natural world is ultimately irrelevant to human economic activity. Near-

perfect substitutability implies that economic output can remain constant or even increase as natural resource inputs are reduced to zero, so long as labor and manufactured capital are increased enough to make up the difference.

These conclusions will seem intuitively unreasonable to most people — and in fact they are. Here's why:

First, if natural and manufactured capital were perfect substitutes for each other, then there would be no need for manufactured capital. If trees grew in stacks as eight-foot two-by-fours, there would be no need for chainsaws, lumber mills, or tape measures. If crude oil could be burned directly in auto engines, we wouldn't need refineries. If steel came in sheets straight out of the ground, then iron smelters, steel mills, and stamping machines would be unnecessary.

Obviously none of these things is true. Manufactured capital is necessary because natural capital is rarely usable in a modern economy in its original, natural form. (Even the flint tools ubiquitous among early human cultures had to be shaped.) The useful end-products result from the complementary interaction of the two forms of capital.

Second, manufactured capital cannot be conjured out of thin air, even by economists. It must be made from something. That something is...natural capital! The steel mills, refineries, and chainsaws are made from ores taken out of the ground by mining machines (more manufactured capital), which are themselves made from ores, and so on. Enabling the entire process are people's accumulated knowhow and skill (human capital).

We saw earlier that production is really consumption of natural resources. Natural capital is turned into manufactured capital, which is used to make more manmade capital and to produce a flow of products for human consumption. The production process cannot totally dispense with either one. This is not to say that one kind of natural capital cannot be substituted for another (as when plastic made from petroleum is substituted for wood or metal). Nor is it to say that the efficiency with which we use natural capital cannot be increased. But, at bottom, there is no escaping the fact that natural and manufactured capital are generally complements (Costanza and Daly, 1992).

It follows that allowing our stocks of natural capital to be overexploited or neglected is a prescription for trouble. Even from the standpoint of natural capital's contribution to the human economic production process, the degradation of natural capital stocks undermines our ability to sustain or increase economic output. Equally important is the fact that natural capital produces not only a flow of *materials* but also a flow of *services* (photosynthesis, water purification, waste absorption, etc.) for which human ingenuity will probably never find comprehensive substitutes.

Degraded natural capital eventually becomes incapable of yielding a flow of materials or these indispensable services. At that point, no human economic activity of any consequence is possible.

Technological Optimism vs. Prudent Skepticism

The machine does not isolate man from the great problems
of nature but plunges him more deeply into them.
—Antoine de Saint-Exupery, *Wind, Sand, and Stars*

Such is our faith in technology that we are strongly tempted to assume it will allow the circumvention of the issues of scale, carrying capacity, and the limits of capital substitutability, and thereby avoid the difficult choices they pose. Technological change seems rapid, inexorable, and good. And at various times and places, technology has allowed evasion of local resource constraints. But having moved from a world in which the human presence was small to one where it is very large, we now face much more fundamental problems, and technology may not offer the answers we need.

To begin with, the effects of new technologies cannot be predicted and the historical record is replete with examples of malign surprises.* The most famous environmental example is chlorofluorocarbons (CFCs). When they were introduced in 1935, CFCs seemed like a chemist's dream: they were inert, nontoxic, nonflammable, noncorrosive, noncarcinogenic, nonmutagenic, incredibly useful, and cheap. No one could have foreseen their destructive effect on stratospheric ozone.

Even when technologies don't turn on their creators, they don't always bring even the gross economic benefits expected of them:

Technologically, the past 15 years have been a parade of wonders: Especially in computation, but also in communication (remember life before the fax?), there has been one revolution after another, with new areas such as biotechnology now seemingly on the verge of widespread practical implementation. Yet

---

* See, for example, Tenner, E., *Why Things Bite Back: Technology and the Revenge of Unintended Consequences*, Knopf: New York, 1996.

economically, the news has been generally dreary—a typical
American worker can buy less with his pay today than his father
could when Richard Nixon was first inaugurated (Krugman,
1990, p. 172).

Apart from the unpredictability of its effects, technology can work for
either good or ill on carrying capacity. The impact on carrying capacity
of human population growth should be expressed in the formula developed
by biologists Paul Ehrlich and John Holdren: I = PAT (impact depends on
population × affluence, or per-capita resource consumption, mediated by
technology). Which technologies are developed and how they are used
determines whether they can be expected to increase or reduce human
impact on the environment. Technologies applied to help extract more
want-satisfaction from a given unit of resource input can increase welfare
without increasing impact. This is using technology in the service of
economic *development* rather than growth (development here meaning the
sense of evolving to a better or more fully realized state).

But technologies can also be used—and predominantly have been
used—simply to extract resources more efficiently and to increase total
throughput (growth). The industrial revolution was a watershed in this
process: an economic Big Bang that rapidly accelerated growth from
relatively low, traditional levels to those we consider normal today (and
the advocacy of which has become one of the chief dogmas of modern
economics). The ability to extract fossil coal, oil, and natural gas, and to
develop machines to use them, greatly increased both the capacity for
economic output and the demand for resource inputs. Much of the historic
improvement in industrial and agricultural productivity, hence economic
growth, has come from increased resource throughput (Daly, 1991b, 1992).
To return to our earlier analogy, we got more power out of our economic
engine mostly by feeding it more fuel and air (resources), not by converting
it to use overhead cams, variable valve timing, and electronic engine
controls (i.e., more efficient technology).

Projections of future increases in productivity generally assume further
increases in throughput. For example, the 1980 *Global 2000 Report to the
President*, prepared by the U.S. Council on Environmental Quality and the
Department of State, assumed that meeting future food production needs
would require doubling or tripling the current inputs of chemical fertilizers,
herbicides, and pesticides (Grant, 1992). Yet if the global ecosystem is
already deeply stressed, as it appears to be, increases of this magnitude
are impossible in the long term.

But what about increasing the efficiency with which we use resources (i.e., development)? How much more welfare can be squeezed out of each unit of resource consumption?

No doubt quite a lot. Energy use provides a good example. Every year, consumers buy millions of inefficient incandescent light bulbs in order to light their houses. Incandescent bulbs typically turn about 90% of the electricity they consume into heat and only 10% into light. If the service (illumination) could be supplied with some other kind of light bulb, many consumers would switch, especially with the proper incentives. In fact, there are super-efficient light bulbs on the market that can save vast amounts of energy. (Getting consumers to buy them is a central feature of many electric utilities' programs to reduce electricity demand and thus avoid building expensive new power plants.) Some analysts argue that the potential for energy savings in other areas of electricity use throughout the economy is equally vast.

Surely other human wants can be satisfied more efficiently as well. And because we have for so long focused on using technology in ways that increase resource throughput, comparatively little work has been done on the possibilities for using technology to increase resource productivity (Daly, 1992). There is a great deal of room for improvement.

Many technological optimists claim that recent events show just that very potential. They argue that the oil shocks of the 1970s triggered a rush toward energy-efficient technologies that effectively "decoupled" energy use and the economy, proving that economic growth need not depend on ever-greater energy inputs after all. And it is true that the energy efficiency of the U.S. economy (expressed as the ratio of gross national product to total fuel consumption) rose 78% between 1929 and 1983; half of that improvement came after 1970 (Gever et al., 1991).

But the nature of this change may have been badly misunderstood. Economic studies carried out at the University of New Hampshire's Complex Systems Research Center suggest that almost all of the apparent improvements in energy efficiency are the result of changes in energy allocation and in fuel-use patterns:

> ...[T]he conventional wisdom overlooks two important structural changes...in the way technology uses energy. One change involves the switch in our primary energy sources from bulky, solid fuels like wood and coal to more efficiently handled fuels like oil, natural gas, and electricity. The other change is that the amount of fuel used in the household sector has declined.... Together, these structural changes have a 96-percent correlation with the year-to-year variation in the ratio of GNP to fuel

consumption since 1929. Little change is left over to attribute to replacing energy with improved machinery or to better manu-facturing processes. ...[P]rices...have had only a small effect on the GNP/fuel ratio. ...We have not, as is usually thought, been doing the same things with less energy. Instead, we have been doing different things with different fuels (Gever et al., 1991, pp. 22-3).

Households use energy for space- and water-heating and for getting around, consumption activities which do not contribute directly to eco-nomic output because they do not produce other goods and services for resale. So diverting fuel from households by getting people to lower their thermostats and drive fewer miles in more fuel-efficient vehicles allows more fuel to be used in industrial activity, where fuel is only one of several inputs. This multiplier effect raises the GNP/fuel ratio. As for the fuel mix's effect, fuels like coal and wood are hard to handle and use efficiently compared to oil and natural gas, as well as less easily and efficiently transported. All things considered, oil and gas produce more economic value per unit of energy.

What effect might these two structural forces have in the future? Continuing to divert fuel from household uses would allow the GNP/energy ratio to continue to improve, but could easily lower standards of living. Increasing the portion of the nation's energy budget supplied by oil and natural gas is difficult or impossible in the long-run, since domestic production is in long-term decline and overseas production will eventually reach the same stage. Barring major and unpredictable breakthroughs in solar energy techologies or a sea-change in public opposition to nuclear power, we may be forced back to more coal and wood. That would decrease overall energy efficiency (Gever et al., 1991).

Another problem with the salvative power of energy efficiency is that it may not actually save energy in the long-run. Economists are not unanimous on this point (see, for example, Lovins, 1998), but many argue that without increases in energy prices, higher efficiencies tend to cause what British economist Horace Herring calls a rebound effect (Pearce, 1998). At the household level, buying a more efficient furnace may encour-age the homeowner to turn up the thermostat. If many households find appliances cheap to run, they may be encouraged to buy more of them. At larger scales, energy efficiency reduces demand for energy resources and thus makes them cheaper, which then leads to greater economic activity and higher energy demand. Herring and other economists argue that claims for the "decoupling" of energy use from economic growth are mostly illusory.

Sobering second looks like this reinforce the suspicion that technology simply cannot bear the weight of our expectations. The world's rich nations expect technology to sustain their endless quest for ever-higher standards of living. The world's poor nations expect technology to make them rich. A simple calculation suggests both will be disappointed.

Assume, for the sake of argument, that the world's rich people are perfectly content with their current standard of living (the high-income countries enjoy per-capita annual incomes of about $25,000). Further assume, quite reasonably, that the world's poor- and middle-income nations aspire to something like the rich nations' standard of living. At the moment, per-capita income in the rich nations is about 23 times that of the poorer nations. For the latter to catch up, without any further impact on the environment, technology would have to improve the efficiency of want-satisfaction per unit of resource consumption by a factor of 23, even if there was no further growth in population. If population actually increases from 5.9 billion in 1998 to 9.4 billion by 2050, as projected (World Resources Institute, 1998), technology would have to improve by a factor of 37 during the period.

Moreover, this assumes that being rich is a static condition, when actually it is a moving target. Few people seem content with a particular level of personal wealth for very long. Our satisfaction seems to depend upon being better off than others or better off now than we were a day, a week, or a year ago. If we permanently succumb to the siren song of affluence, our wants will be insatiable. In that case, even perpetual technological improvements in efficiency will never be enough.

The debate about the possibilities for technology-driven economic growth and development is long-standing, mainly because uncertainty is high. Only time will tell whether the technological optimists are right. But what to do in the meantime?

It is useful here to recall an important lesson of the history of science. Human knowledge, according to chemist Henry Bauer, can be divided into three categories. The *known* is the conventional wisdom, or what we believe to be true. The *known unknown* is the knowledge that the known implies and can be expected to be found through further inquiry. For example, it appears that microorganisms cause disease. When a new disease appears, the first efforts to understand and treat it focus on identifying the previously unknown organism that is expected to be there. The third category is the *unknown unknown*, and it regularly produces shocks, partly because scientists periodically come to believe they have explained most of the important facts of reality and have only details left to fill in:

> The conventional wisdom is blind to its own inadequacy, to the fact that, sooner or later, it will be found to be wrong, in one way or another. Each generation of scientists has believed that it understands...the chief principles that govern natural phenomena. Following Newton, many generations were convinced that all physical phenomena are just matters of particles and forces. ...By about 1870, scientists felt quite secure about the main principles with which all phenomena could be explained. But within a few decades, entirely out of the blue, came radioactivity, the discovery that some atoms self-destruct; and then the necessity to describe radiation sometimes as particles...rather than as waves; and then relativity, non-Euclidean space, the uncertainty principle: a succession of total surprises.... ...Human beings, including scientists, do not function under continual awareness of humanity's fundamental ignorance; rather, they live under perpetual illusion of fundamental understanding (Bauer, 1992, pp. 74-5).

To proceed recklessly, buoyed by technological optimism, is to flout the risks lurking in both the known unknown and the unknown unknown of ecology. To do so ignores the rule that says, when in doubt, choose the option that leaves the most options later. Failing to leave such options open puts both ourselves and our descendants at risk; as someone once said, it is immoral to pay tribute to the technological prowess of our descendants by guaranteeing that they need it. If we continue along the present course and hope for the best, we risk destruction of the resource base and irreversible environmental degradation. Prudence therefore calls for action to reduce the severe stresses created or threatened by overconsumption among the rich and rising population and the hunger for ever-increasing affluence among the poor. Prudence dictates aiming for sustainability. If that policy turns out to be unnecessarily conservative, we can leave our children and grandchildren an endowment in the form of an extravagantly—not just adequately—healthy, beautiful, spacious, and livable world.

Entropy and Economics

> Things fall apart; the center cannot hold;
> Mere anarchy is loosed upon the world....
>   — W.B. Yeats, *The Second Coming*

Things fall apart because it is the law — the second law of thermody-namics. (The first law of thermodynamics is the law of conservation, which says that matter and energy cannot be created or destroyed, only trans-formed. Matter is itself a form of energy, as is shown by Einstein's famous equation, $E=mc^2$.)

The second law of thermodynamics was developed in connection with steam engines in 1824 by French physicist Nicholas Léonard Sadi Carnot. Carnot realized that using energy to do work (move matter through space) depended on the machine's temperature gradient, i.e., the difference between the hottest and coolest parts. As the work is performed, it reduces the temperature differences. Although the energy total remains constant, it becomes less available to do further work (Boulding, 1981a).

More generally, using energy makes it less available. The latent chem-ical energy in fireplace logs is highly available until it is released by burning. Thereafter, although the amount of energy in the heat, gases, and ashes is the same as the amount that was in the wood, it is scattered and thus less available. In theory, it is possible to reassemble the com-ponents and reconcentrate the energy, but doing so would take more energy than it would yield (Daly and Cobb, 1989).

Another way to express the entropy law is that objects and systems tend to disintegrate over time. They break, break down, break up, rust, die, decay, wear out, or generally move from a state of higher organization to one of lower organization, from order to disorder. As far as is known, this process always moves in the same direction. (Irritatingly, since entropy is a measure of the disorder in a system, a highly organized system is said to be low-entropy, while a disordered system is said to be high-entropy. Entropy increases as order decreases.) The breaking down and wearing out of a system or object can be stopped if it is an *open* system capable of receiving inputs of matter and energy — maintenance — from outside. Even a *closed* system, which allows only inputs and outputs of energy, can maintain order over time. In an *isolated* system (one in which there is no traffic in inputs or outputs, i.e., no throughput), disorder must increase.

Is life an exception? Doesn't life create order out of disorder? Anyone with small children will immediately doubt this. Yet life in general appears to be an example of movement from a state of lower organization to one of higher organization. After birth, human beings gradually grow out of utter helplessness to relative independence, learning to survive in the world and do things of extraordinary complexity, up to and including writing novels, symphonies, and arcane mathematical tracts on things like entropy. Doesn't evolution in general, with its vast, eons-long procession of movement from creatures of one-celled simplicity to blue whales, prove that entropy can be beaten?

Yes and no. On a local scale, yes: life has indeed evolved marvels of increasing organizational complexity. But in terms of the big picture, no. Living creatures exist only by being able to "import" highly complex, low-entropy matter (i.e., to eat food), extract useful energy and materials from it, and "export" wastes of much lower complexity (higher entropy). All life on Earth recycles itself through the ecosphere in this manner, each creature using something from its surroundings (usually including other creatures) to sustain and recreate itself. Matter is not created or destroyed, only broken apart and reassembled to be used again in some other form. As physicist Erwin Schrödinger once put it, life (and evolution) can be seen as the segregation of entropy; "[T]he device by which an organism maintains itself stationary at a fairly high level of orderliness (= a fairly low level of entropy) really consists in continually sucking orderliness from its environment" (Schrödinger, 1967, p. 79). Life creates pockets of order at the cost of disorder elsewhere. Evolution is pollution (Boulding, 1981a,b).

Humans and other living things are thus clearly open systems. However, the biosphere and the Earth itself are closed systems. Matter is essentially constant; little comes into the system except the occasional stray chunk of comet or meteorite, and little goes out except space probes. But in terms of energy, the flow of solar radiation coming in (balanced by the flow of re-radiated heat) is continuous and crucial. It is the ultimate answer to the question, "How does the economy (and the world) work?" Hazel Henderson (1981) tells of a paper delivered by English Nobelist Frederick Soddy in 1921 in which he used the steam locomotive as a metaphor, asking "What makes it go:"

> In one sense or another the credit for the achievement may be claimed by the so-called engine-driver, the guard, the signalman, the manager, the capitalist, or the shareholder — or, again, by the scientific pioneers who discovered the nature of fire, by the inventors who harnessed it, by Labor, which built the railway

and the train. The fact remains that all of them by their united efforts could not drive the train. *The real engine driver is the coal.* So, in the present state of science, the answer to the question how men live, or how anything lives, or how inanimate nature lives, in the senses in which we speak of the life of a waterfall or of any other manifestation of continued liveliness, is, with few and unimportant exceptions, BY SUNSHINE (p. 225).

"Needless to say," Henderson writes, "Soddy was considered a crank." But he was right: the steady imports of solar energy drive the life processes of Earth. If the Earth were closed to the solar flow, which is low-entropy energy made generally available to the biosphere through photosynthesis, all life would eventually cease. Of course, the sun is not exempt from the entropy law either; it is slowly running down as it burns up its nuclear fuel and will come apart, spectacularly, in a few billion years.

What has entropy got to do with economics?

The laws of thermodynamics are relevant to the economy because economic activity is entropic. Natural resources (low-entropy matter-energy) are gathered, processed to separate out the useful parts from the rest, manufactured into goods, and transported to the point of sale. Wastes are produced and energy is used up (and made less available) every step of the way. The quantity of raw materials is equal to the quantity of wastes (plus the products, which eventually become wastes), but the two amounts are *qualitatively* different. The difference is measured in terms of entropy. Economic production is utterly dependent on the availability of low-entropy inputs (Daly and Cobb, 1989).

These inputs come from two sources. As noted above, one is the sun. The other is the Earth, which yields useful minerals, plant and animal life, and fossil fuels. There are obvious differences between the nature of the solar and terrestrial inputs, but perhaps even more important is the radical difference in their availability. The stocks of solar inputs are essentially infinite but the flow is limited and, for the present, relatively unavailable directly to the economy. (Solar energy drives climatic processes and the ecosystems that provide essential economic services, but we use little solar energy directly to run machines or heat buildings.) As for earthly inputs, stocks of renewable resources are limited by the available flow of solar energy. Earthly stocks of nonrenewable resources (especially fossil fuels, which are really deposits of solar energy laid down millions of years ago) are finite, but by means of technology we can extract them from the ground and pour them into the economy at enormous rates (Daly and Cobb, 1989).

Economic activity relentlessly degrades the stocks of both renewable and nonrenewable resources. Fortunately, inputs of low-entropy energy allow them to be restored. The solar flow can continually renew the stocks of renewable resouces, provided we don't interfere with them too much. Degraded stocks of many nonrenewable resources (such as ores that have been extracted, made into products, then dissipated through rust and decay—though not fossil fuels, obviously) can theoretically also be reconstituted if enough useful energy is available. The catch for our current economic system is that, in the long term, using fossil energy to do this is unsustainable because of the effect on climate and ecosystem integrity. However, the amount of solar energy falling on the Earth is thousands of times greater than our current needs (Ayres, 1998). If technologies were developed to tap that energy efficiently, we could even recover the dispersed mineral ores dissolved in sea water.

Technology has enabled the human economy to temporarily suspend its dependence on the solar source of low-entropy inputs—which is what marked the threshold between pre-industrial and industrial society. Cheap fossil fuels have, in effect, kept the solar flow limited by making it uneconomic to develop efficient means of tapping it. But technology cannot abolish the entropy law, and it will have to find a way to make much better use of the low-entropy solar flow that is ultimately the only large-scale sustainable source of energy.

What Sustainability Means

ENI for Sustainable Development: We're Growing With the Planet.
> — Italian energy company slogan

The best-known definition of sustainability was authored by the Brundtland Commission in its 1987 report and goes something like this: sustainable development means ensuring our ability to meet the needs of the present without compromising the ability of future generations to meet their own needs. The commission's solution to this challenge was to call for more economic growth of an order we have already argued is not possible without fatal damage to the environment. However, even if its solution is unworkable, the Brundtland Commission started something big. Although it has brought forth a share of absurd slogans like the one above, sustainability has become at least a nominal goal for policymakers worldwide. The commission deserves enormous credit for its courage in bringing the issue into clear and unavoidable view and launching a search for consensus on what sustainability requires.

Ecological economists believe that close attention must be paid to the critical role natural capital plays in sustaining the resource flows and ecosystem life-support services that human beings and other creatures rely upon. Our use of these benefits creates a wide range of future risks with inherently unknowable probabilities. This is a compelling argument for an approach that conserves as many options as possible, both for ourselves and for the generations to come. In our present state of knowledge, or ignorance, we simply cannot know what we can safely do without.

A sustainable system is one with sustainable income. A good definition of income in this context is that offered by economist Sir John Hicks in 1948: "The purpose of income calculations in practical affairs is to give people an indication of the amount which they can consume without impoverishing themselves. Following out this idea, it would seem that we ought to define a man's income as the maximum value which he can consume during a week, and still expect to be as well off at the end of the week as he was at the beginning" (cited in Daly and Cobb, 1989, p. 70).

We have already seen that what yields income (flows) is capital (stocks). No matter whether the capital in question is manufactured or natural, by Hicks' definition one cannot consume more than the flow provided by the stock and still remain as well-off. Consuming the stock too means that present wealth and future income have both declined. So the most important implication of the Hicksian definition of income is that *the stock (capital) must be kept intact* (Daly and Cobb, 1989).

Preserving our capital intact is the minimum safe condition for achieving sustainability. We need both manufactured and natural capital to do it. There are two major conceptions of sustainability, differing in their treatment of the relationship between manufactured and natural capital. *Weak sustainability* calls for maintaining the combined total of both natural and manufactured capital intact. This approach suggests that we could allow natural capital stocks to decline, provided manufactured capital stocks were rigorously built up to compensate. Even this would improve on present practice. However, in view of the evidence that natural and manufactured capital are largely complementary, the wiser course would be to pursue a policy of *strong sustainability*, which calls for independently maintaining the stocks of both manufactured capital and natural capital.

In any economic scenario, production is limited by the least available factor. The growth of manufactured capital and population, coupled with the evidence of environmental degradation already available, suggests that the scarcest factor now is natural capital. With human economic activity continuing to draw down the Earth's stocks of natural capital, our very first priority should be to halt its further depletion as soon as possible. We should next begin immediately to invest in natural capital so as to restore it to noncritical levels. And in general we need to adopt the following three criteria for conducting economic activity so as to maintain natural capital and ecological sustainability:

1. For renewable resources (fish, trees, etc.), the rate of harvest should not exceed the rate of regeneration.

2. The rate at which we allow economic activity to generate wastes that must be passed into the environment should not be allowed to exceed the environment's ability to absorb them.

3. The depletion of nonrenewable resources (oil, coal, etc.) should be offset by investment in and development of renewable substitutes for them.

Sections II and III discuss in more detail the nature and functions of natural capital and what is required to manage the world's stocks of natural capital for long-term sustainability.

THE DEFINITION, FUNCTION, AND VALUATION OF NATURAL CAPITAL

Chapter 3

What Natural Capital Is and Does

A tree is a tree — how many more do you need to look at?
— Ronald Reagan

The concept of natural capital is an extension of the traditional economic notion of capital, which is generally defined as the manufactured (human-made) means of production, i.e., machinery, tools, equipment, buildings, and so on. What natural capital and manufactured capital have in common is that they both conform to the working definition of capital as a stock (collection, aggregate) of something that produces a flow (a periodic yield) of valuable goods or services. A stock of factory machinery produces a flow of clothing or automobiles. A stock of trees (i.e., a forest) produces a flow of goods in the form of new trees *and a flow of services* in the form of oxygen, erosion control, wildlife habitat, etc. As noted earlier, we can count the flows as income. Depleting the stocks, however, is called capital consumption.

Natural capital can be divided into two major categories and a hybrid one. The major categories are renewable natural capital and nonrenewable natural capital. Renewable natural capital is living and active. Ecosystems consist largely of renewable natural capital. Renewable natural capital continuously maintains and regenerates itself (when left alone) by harvesting solar energy and converting it through photosynthesis to plant mass and thus, eventually, into the rest of the food web. As long as it is intact and the sun shines, renewable natural capital will yield a steady flow of useful goods and services. In that sense, it is essentially unlimited. The volume of the flow, however, is finite. Overuse of renewable natural capital can impair or destroy its ability to regenerate itself and sustain the flow of goods and services on which we depend.

Nonrenewable natural capital (mostly fossil-fuel and mineral deposits), on the other hand, is passive. The Earth's stocks of nonrenewable natural capital are finite, but the flow rate is pretty much a matter of policy; there's only so much oil in the ground, but we can pump it out rapidly or slowly, as we choose. Nonrenewable natural capital yields essentially no services until it is taken out of the ground and converted into some useful form (Costanza and Daly, 1992).

The hybrid category might be called cultivated natural capital (Daly, 1994). This includes all agricultural and aquacultural systems, such as tree farms, sod farms, fish ponds, and greenhouse nurseries. The components of these systems are not manufactured by humans, but they're not entirely natural either. Humans create cultivated natural capital by taking elements of natural capital and changing the way they function through selective breeding, use of monocultures, and so on.

Cultivated natural capital combines features of natural capital and manufactured capital. A tree farm, for example, is essentially a streamlined forest; the trees themselves are natural capital, but they have been rearranged into a configuration and a kind of biological community not found in nature. It is the manufactured capital services — such as planting in close, orderly ranks and the culling of diseased trees — that give the plantation its distinctive character. Densely planted tree farms are often dark and gloomy and tend to lack the understory vegetation of natural forests. The benefit of higher-than-natural timber production is purchased at the cost of degraded wildlife habitat, reduced biodiversity, and greater vulnerability to pests and disease. In this way, cultivated natural capital can substitute for natural capital in some ways but not others.

Besides manufactured capital and natural capital, there is a third form, called human capital. Like manufactured capital, it too originates with people. Human capital is the collective knowledge, skills, and culture of the species. Human capital (sometimes called cultural capital) provides the means by which people not only respond and adapt to the natural environment but also modify it for their own purposes. It encompasses people's views of the natural world and the systems of ethics by which they decide what actions ought and ought not be taken with respect to the natural world. It also includes the philosophies and cosmologies upon which those ethical systems rest, and the social institutions people construct to manage their actions toward the natural world. Finally, human capital includes scientifically accumulated data and theories as well as local, personal, and traditional knowledge of the environment and its functioning (Berkes and Folke, 1994).

A logger notching a redwood with a chainsaw is thus not merely performing an act of labor. The logger embodies his culture's values and knowledge, and his act takes place at the confluence of human capital and natural capital. His act encompasses the knowledge and skill necessary to conceive, design, and construct a chainsaw and use it properly and efficiently, the knowledge of which trees are best to cut for which purposes, the social and political arrangements by which someone is allowed to cut a particular tree in a particular place, the attitude with which the tree is cut, and the understanding of (and concern about) the effects — on soil erosion, stream siltation, wildlife diversity, and so on — of harvesting the tree. The human capital dimension is a critical element in the health of the ecosphere; as we saw in Chapter 1, it was an important transformation of human capital, in the form of changes in social institutions and prevailing attitudes toward the natural world, that set the stage for the current environmental drama.

Summing up, capital comes in three major forms: natural, human, and manufactured, corresponding to the conventional production-factor trio of land, labor, and capital. All of them decay, by virtue of the entropy law. The rate of decay for nonrenewable natural capital is so slow that we can disregard it, but the other types must all be maintained to be useful in the long term.

What Does Natural Capital Do?

The materials of wealth are in the earth, in the seas, and in their natural and unaided productions.
— Daniel Webster

There are many ways to categorize the roles of various kinds of natural capital. One is to describe natural assets in terms of their functional contribution to regulation of the biosphere, production of food and other useful items, habitat, or information (see sidebar). Another scheme categorizes natural capital in terms of its direct and indirect economic value (see "Biodiversity, Ecosystem Function, and the Economy" to follow).

Regardless of which classification scheme is used, viewing natural capital as merely natural resources is a blinkered view of its role and function. *The primary value of natural capital is life support.* All the rest is secondary.

Natural capital as a life-support system provides the biophysical necessities of life. In effect, life makes life possible. The ecosphere consists of the organisms, processes, and resources that interact to provide food, energy, mineral nutrients, air, and water (see sidebar for a representative list).

At the local, regional, and global levels, the functionality of ecosystems ultimately depends on their integrity, or the intactness of the complicated web of interconnections that link species with one another. Every child who has seen the wings pulled off a fly understands that no living system can be completely disassembled and still function. Because the dynamism of a living system derives from its wholeness, biodiversity — the numbers and distribution of living species — is central to the primary life-support value of natural capital.

This is not to say that it is not possible to subtract bits and pieces here and there, to perturb the dynamic system that is the ecosphere to some extent, without serious losses of function. It is somewhat like a toy gyroscope, which can be tapped lightly while it is spinning without causing more than a slight wobble. The gyroscope's spin (its dynamism) gives it a measure of stability. Some scientists also believe that complex systems can continue to function following perturbation better than simple ones, at least in some cases. Put another way, complexity can mean robustness. This may help explain why we have been able to stress and co-opt the global ecosystem, which is the most complex system we have direct experience of, to such a great extent without noticing anything going wrong — and why our appropriation of net primary product (see Chapter 2) is of such concern.

But it is crucial to understand that: (1) the perturbations cannot continue indefinitely; and (2) no one really knows how far the process can go without pushing the level of ecospheric integrity below some critical threshold, to the point at which indispensable life-support functions might degrade or even collapse. It is also important to note that some pieces of ecosystems are more critical than others. Just as a person can survive without a leg but not without a head, an ecosystem may depend more heavily on a keystone species* for functional integrity than on the other species found within it. (One of the most pressing research needs is to discover which are the keystone species.) Biodiversity, because it imparts strength or robustness to the global ecosystem, especially over the long term, is thus one of the most important dimensions of natural capital.

Biodiversity, Ecosystem Function, and the Economy

Biodiversity is the tool with which you play the game of promoting global stability.

— Peter H. Raven

Biologist E.O. Wilson argues that there are really only two kinds of environmental problems (Wilson, 1992). The first kind involves the gross threats to the livability of the planet posed by human economic activity, such as air and water pollution, the depletion of stratospheric ozone, the possibility of greenhouse warming, and the rest. The second kind — biodiversity loss — is a consequence of the first kind. Environmental degradation and human overuse of natural ecosystems lead quickly to the destruction of species at rates that greatly exceed the natural background rate. The first kind of environmental problem is reversible in principle, but no one knows how to bring back lost species, much less the ecosystems they constitute.

Why is the loss of biodiversity bad?

One response is that it isn't, at least not entirely. It certainly is not unnatural. Various estimates put the natural background rate of extinction at one mammal species every 400 years and one bird species every 200 years (World Resources Institute, 1994), or, more broadly, one out of every million species every year (Wilson, 1992). These estimates apply only in biologically placid times; in the last 500 million years there have been at least five catastrophic extinction episodes (including the one that ended the reign of the dinosaurs 65 million years ago) and a host of more minor ones. Nature devours her progeny all the time.

---

* A keystone species is one that "affects the survival and abundance of many other species in the community in which it lives. Its removal or addition results in a relatively significant shift in the composition of the community and sometimes even in the physical structure of the environment" (Wilson, 1992, p. 401).

Four Functions of Natural Capital

One way of classifying natural capital and its roles is in terms of four categories of functions: regulation, production, carrier, and information (see Table 3.1) (De Groot, 1994).

The list of regulation functions begins with protecting the Earth and its inhabitants from outside dangers, such as meteorites (most of which are absorbed by the atmospheric blanket) and harmful ultraviolet radiation (much of which is stopped by stratospheric ozone before it reaches the surface). It includes the capacity of natural and human-modified ecosystems to regulate essential ecological processes such as climate, to provide clean air, water and soil, and to store and recycle wastes.

Natural capital's carrier functions are basically those by which it provides humans and other creatures with suitable places to live and make a living, biologically speaking. Production functions include resources such as oxygen, water, food, fuel, medicines, biochemicals, and so on. Genetic resources are included here too, since the genetic library represented by the myriad species living on Earth forms the basis of our agriculture. Finally, natural capital's information functions include the opportunities the natural world provides for learning, artistic inspiration, and aesthetic and spiritual experience.

By way of illustration, consider the natural capital functions of tropical moist forests (rain forests), such as those found in parts of Central and South America:

Regulation. Tropical moist forests cool the land by shading it and by evaptranspiration, so that clearing a forest tends to raise the temperature of the land. This can create thermals (currents of rising air) that disperse rain clouds, thus making rainfall over cleared lands more erratic and less frequent, or even causing drought. Forests regulate watersheds by holding water, thus preventing flooding and controlling water flow from hillsides to irrigation systems in the valleys below and to navigable streams and rivers. Forests are tireless and prolific photosynthesizers, converting solar energy to forms indispensable to people and other creatures. Tropical moist forests, many of which are millions of years old as ecosystems, are richly biodiverse and support many complex relationships among various species. These relationships often create important biological control mechanisms that promote pollination and control pests.

Carrier. Despite high rates of destruction, tropical moist forests still provide places to live for many indigenous peoples. The relationships these peoples have with their habitats also shape their cultural identities.

Production. Forests are major sources of timber, which could, in principle, be harvested sustainably. Apart from timber, about one species in every six living in tropical moist forests is believed to be useful. These useful items include foods such as fruits and nuts. Forests also harbor many species that are now cultivated artificially for foods and raw materials. The cultivated varieties require periodic genetic rejuvenation to improve productivity or appeal and to develop new pest- and

continued on p. 56

Table 3.1   Functions of the Natural Environment

Regulation Functions	
1.	Protection against harmful cosmic influences
2.	Regulation of the local and global energy balance
3.	Regulation of the chemical composition of the atmosphere
4.	Regulation of the chemical composition of the oceans
5.	Regulation of the local and global climate
6.	Regulation of runoff and flood prevention (watershed protection)
7.	Water catchment and groundwater recharge
8.	Prevention of soil erosion and sediment control
9.	Formation of topsoil and maintenance of soil fertility
10.	Fixation of solar energy and biomass production
11.	Storage and recycling of organic matter
12.	Storage and recycling of nutrients
13.	Storage and recycling of human waste
14.	Regulation of biological control mechanisms
15.	Maintenance of migration and nursery habitats
16.	Maintenance of biological (and genetic) diversity
Carrier Function—providing space and a suitable substrate for:	
1.	Human habitation and (indigenous) settlements
2.	Cultivation (crop growing, animal husbandry, aquaculture)
3.	Energy conversion
4.	Recreation and tourism
5.	Nature protection
Production Functions	
1.	Oxygen
2.	Water (for drinking, irrigation, industry, etc.)
3.	Food and nutritious drinks
4.	Genetic resources
5.	Medicinal resources
6.	Raw materials for clothing and household fabrics
7.	Raw materials for building, construction and industrial use
8.	Biochemicals (other than fuel and medicines)
9.	Fuel and energy
10.	Fodder and fertilizer
Information Functions	
1.	Aesthetic information
2.	Spiritual and religious information
3.	Historic information (heritage value)
4.	Cultural and artistic inspiration
5.	Scientific and educational information

continued from p. 54

disease-resistant strains. About one-quarter of tropical forest species are useful as medicines, and in fact roughly 20% of the billions of prescriptions written each year are for drugs that originated in forest species. Many other species provide useful biochemical or biodynamic compounds, such as latex.

Information. Forests offer less tangible benefits as well, in the form of perceptions and experiences that are the raw material for education, art, and spiritual development. For non-native peoples, forests are increasingly providing opportunities for recreation and tourism.

However, except for the rare asteroid impact or planet-wide episode of vulcanism, nothing wipes out species faster than human economic activity, from Neolithic hunter-gatherers decimating mastodon herds to modern developers wiping out habitats for housing tracts and golf courses. But therein lies the other argument against the unmitigated badness of biodiversity loss: it has been necessary for much human economic progress. As humans have struggled to make better lives for themselves over the millenia, their progress has depended heavily on learning to use certain natural resources, such as domesticatable animals and plants, in specialized and intensified ways. This specialization and the sheer extravagance of human success have crowded out many other species. Further losses of biodiversity may be both necessary and inevitable:

> [A] decision to preserve the biotic status quo may very well condemn future generations [of humans] to progressive impoverishment, especially in light of the continuing expansion of human population. ...[I]t may very well be that growth is a necessary condition for the protection of biodiversity in the future. ...[W]ithout growth in the personal incomes of the poorest two-thirds of the world's inhabitants, and without a demographic transition that depends on growth in personal incomes, there is little hope of halting the march into the world's remaining forested areas (Perrings, 1994, p. 93).

Nevertheless, tolerating too much loss of biodiversity would be a grave mistake. Although much is unknown about the quite complex relationship between ecosystem function and biodiversity, broadly speaking the former depends on the latter. Loss of biodiversity can erode ecosystem function, at least in particular times and places and possibly on larger scales as well. Therefore, if we wish to err on the safe side in preserving ecosystem function, especially over time scales of decades or centuries, we must preserve biodiversity.

This constraint clearly does not apply to the present generation alone. We have argued that perpetual economic growth is not possible. If that is so, then the only hope for continuing improvement of human welfare is sustainable development. But that condition cannot be satisfied if the value of natural capital, including biodiversity, is allowed to decline:

> [T]he economic and ethical problems converge in the need to maintain that level of biodiversity which will guarantee the resilience of the ecosystems on which human consumption and production depend. ...[T]he weight of available evidence suggests that the loss of biotic diversity...is compromising the interests of future generations precisely because it is narrowing the range of options open to those generations. There is no evidence that any care is being taken to assure that the accelerating destruction of biota by the present generation will not limit the range of choice open to future generations. Biota are being destroyed to maintain consumption, not to promote investment (Perrings, 1994, p. 110).

The danger of permitting heavy losses of biological diversity becomes obvious by examining its nature and its close links to both ecosystem function and the human economy.

Biodiversity is usually taken to mean *species diversity*, i.e., the numbers and distribution of species and populations of species. But biodiversity also has at least three other key dimensions (Jansson and Jansson, 1994). Perhaps the most important is *functional diversity*, which means what the different species in an ecosystem do, or what their biological jobs are. Functional diversity is important because it appears to be a critical variable in both ecosystem productivity and stability.

To a considerable extent, productivity varies directly with diversity:

> Field studies show that as biodiversity is reduced, so is the quality of the services provided by ecosystems.* Records of

---

* Experimental evidence has begun to emerge that supports the conclusions of the field studies. At least in experimental ecosystems with relatively low diversity, adding species that increase the functional diversity of the system also increases its efficiency at converting carbon dioxide into plant tissue, thus leading to greater productivity (Naeem et al., 1994).

stressed ecosystems also demonstrate that the descent can be unpredictably abrupt. As extinction spreads, some of the lost forms prove to be keystone species, whose disappearance brings down other species and triggers a ripple effect through the demographies of the survivors. The loss of a keystone species is like a drill accidentally striking a powerline. It causes lights to go out all over (Wilson, 1992, pp. 347-8).

It is important to note, however, that the relationship between diversity and productivity is not strictly linear, i.e., the latter does not always increase with the former. In plant communities, species diversity is typically highest where the soil is, at best, only middling fertile and productivity is thus only intermediate.* Some highly productive ecosystems are fairly simple and require relatively few species to function. Estuaries, which may often harbor a few very flexible species that are able to feed on many different foods as circumstances change, are among the most productive ecosystems anywhere (Costanza et al., 1993).

The relationship of biodiversity to ecosystem stability is more complex and contentious among ecologists. For one thing, there are several different definitions of stability (Holling, 1973). In one view, stability means resilience: the ability of an ecosystem to recover when perturbed. As we noted earlier, some ecologists argue that higher biodiversity imparts greater resilience, so that when an ecosystem harbors many different species performing the same functions — such as photosynthesis, nitrogen fixation (making nitrogen available to plants, a job performed by certain bacteria) or decomposition — the system is more likely to survive even when one or more species is lost, because the function can still be carried out by other species. Species-poor ecosystems may lack this redundancy and are thus more likely to be vulnerable to perturbation, even if the particular species involved are individually hardy. However, the progressive loss of species can overwhelm the resiliency of even highly biodiverse ecosystems, as the efficiency of the operation of the food web declines and nutrient flows are attenuated. The loss of a keystone species can lead to the ecosystem's collapse and even change its physical structure.

---

* Tropical rainforests display great diversity but thrive on poor soils. Farmers learn this firsthand when they clear the forest only to find that the soil will only support crops for a few seasons. An important implication of this relationship is that preserving biodiversity need not require setting aside good agricultural land, since the best soils tend to support plant communities of lower diversity (Huston, 1993).

Another school of ecological thought argues that while stability can lead to biodiversity, the reverse is not necessarily true. That is, ecosystems that have been fairly stable for very long periods of time, such as rainforests, may show enormous diversity — but a severe, extensive perturbation, such as clear-cutting, will not necessarily be followed by rapid regeneration and recovery of the system. On the other hand, ecosystems that have evolved in parallel with periodic episodes of creative destruction may be among the most resilient and most capable of self-renewal (Costanza et al., 1993).

To some extent, the argument is driven partly by focuses on different spatial and temporal scales. At small scales and over short periods, ecosystems may function well with a relatively few species that are well suited to local conditions. Over the long term and at the regional or global level, the extent of biodiversity probably becomes more critical. Over those scales, the likelihood of significant perturbation or changes in conditions increases from one locale or era to another. Diversity provides a kind of insurance that, as conditions change and some species decline, a pool of other species will be available to fill the gap in essential ecosystem services.

Bottom line, it is true that there is little or no hard evidence permitting accurate judgments about the number of species that may safely be lost before the ecosystem services upon which the human economy depends are threatened. But no one argues that humans can survive without any other living creatures.

In addition to functional diversity, ecosystems also exhibit *temporal diversity* and *spatial diversity*. Temporal diversity refers to the changes in ecosystem membership and structure that accompany periodic variations in energy and matter flows, such as those triggered by daily and seasonal changes in sunlight. Spatial diversity means the distribution in space of the various job-holders in an ecosystem. For example, in the open ocean the primary producers are concentrated in the photic zone, defined as the topmost layer of the water through which sunlight can penetrate. Depending on the water's turbidity (cloudiness), this zone varies from about a meter to perhaps 200 meters in depth. The chief decomposers, however, live several thousand meters down. The two zones are connected as an ecosystem by the settling of organic waste through the water column, by upwelling currents bearing inorganic nutrients, and by the vertical migration of fish and crustaceans. Forests display a similar vertical stratification, but the distances involved are much shorter. Primary production takes place in the leaves of the tree crowns, while respiration takes place in the root zone, no more than a few dozen meters distant. Transport of nutrients takes place through the tree trunks. In coral reefs, which are among the most efficient recycling ecosystems, producers and consumers in the community are even closer.

Given these dimensions of biodiversity, what are the grounds for valuing it? Our focus here is on its purely practical importance, but in passing we should acknowledge that for many people the nonmaterial worth of biodiversity is uppermost. They believe that all living creatures are beautiful and intrinsically worthy of appreciation and preservation. Others believe that simple ethics prohibits the casual destruction of any living thing; life is extraordinary and quite rare in the universe, for all we know, and so all forms of it are intrinsically invaluable. It has also been hypothesized that human beings have a genetically based tendency to be interested in and engaged with the natural world, a tendency that confers an evolutionary advantage.* If true, the wish to preserve biodiversity is simply human nature.

The practical importance of biodiversity can be described in terms of direct and indirect economic benefits (Ehrlich and Ehrlich, 1991). The direct benefits are the most obvious:

- *Food.* We harvest food from natural ecosystems and from agriculturally modified ones. We also draw huge quantities of protein from the oceans and from coastal wetlands (which both nurture marine life we take directly and feed ocean creatures we consume later).
- *Commodities.* Direct benefits also include timber and other plant products such as rubber, oils, and organic chemicals. Plants that can be used directly as sources of energy (such as in whole-tree-fired electrical generation) or to make biofuels (such as corn and sugar cane) are heavily cultivated in places like Brazil and to a lesser extent in the United States, and are likely to increase in importance in the coming years.
- *Drugs.* In the United States, more than 40% of all prescriptions are for drugs based on natural organisms: about 25% are derived from plants, 13% from microorganisms, and 3% from animals. These include such exotic drugs as anticoagulants derived from leeches and drugs to unclog blocked arteries that come from vampire bats, as well as a host of more familiar substances, including atropine, codeine, digitoxin, L-Dopa, menthol, morphine, penicillin, quinine, reserpine, scopolamine, taxol, and others (Wilson, 1992).

---

* See *Biophilia*, by E.O. Wilson, and *The Biophilia Hypothesis*, by Stephen R. Kellert and E.O. Wilson.

For what it is worth, all of the world's most widely used recreational drugs (licit or illicit), including alcohol, tobacco, coffee, tea, khat, marijuana, cocaine (coca), opium, and a number of hallucinogens, are plants or plant-based.

- ***The genetic library.*** The volume of genetic information contained in the Earth's species is immense. We have drawn heavily on it in the past for our own benefit and continue to do so. Corn, wheat, rice, oats, barley, rye, and sorghum began as native wild grasses that were developed into specialized, productive crops. The domestic animals so integral to human culture — cattle, oxen, horses, sheep, goats, llamas, alpacas, camels, pigs, chickens, and the rest — were all borrowed from the natural library and genetically edited to suit human needs.

The intactness of the world's genetic stores confers at least two important and related benefits. The first is the option of genetic rejuvenation. In borrowing strains of plants and animals from nature and then breeding them to our liking, we have increased crop productivity and otherwise made the creatures greater assets to the human economy. But there is a tradeoff: selectively changing a species to improve it in one or two narrow dimensions tends to reduce its hardiness or create other infirmities. A good illustration of this effect can be seen in the history of thoroughbred horses. Intensively bred since the 18th century for speed over short distances, thoroughbreds have gotten faster and faster at the price of a tendency toward respiratory bleeding, reproductive difficulties, aggression, and even psychotic behavior (d'Arge, 1994). The same kind of tradeoff is evident in plant species bred for duty as crops; the genetically pure strains (generally planted as monocultures) raise yields but often lack resistance to disease and pests and may be unable to survive in their original habitats. They also tend to require coddling in the form of higher inputs of fertilizers and protection from natural enemies by means of pesticides.

In short, homogeneity can mean vulnerability (Wilson, 1992). The untamed relatives of our domesticated species preserve important genetic resources for restoring their domestic cousins' resistance. The Earth's existing genetic diversity represents both a toolbox and a warehouse of spare parts to fix ecosystem problems. Since we cannot predict what will go wrong, we should not discard any tool or part we might need.

The second benefit of the genetic library is the great untapped potential for productive new crops, commodities, and medicines conserved there. Humans have cultivated or collected about 7,000 species of plants for food, but 90% of the world's food supply comes from only 20 kinds of plants, and half of the supply consists of wheat, corn, and rice. Residents of temperate zones enjoy only about a dozen varieties of fruit (peaches, strawberries, apples, and so on). The tropics, however, boast more than 3,000 varieties; 200 of them, with exotic and delectable names like carambola, lulo, rambutan, and durian, are in wide use. Worldwide, as many as 30,000 species of plants produce edible parts. As for drugs, the pharmacological treasurehouse alluded to above represents a tiny fraction of the drugs that might be awaiting discovery; for example, only 3% of the estimated 220,000 species of flowering plants has been examined for useful alkaloids (Wilson, 1992).

If the direct economic benefits of biodiversity seem like commodities, the indirect economic benefits are more like infrastructure. Here are some of the known ones (Ehrlich and Ehrlich, 1991):

- **Photosynthesis.** As mentioned in Chapter 2, primary producers, which convert sunlight to biomass through photosynthesis, are the basis of life on Earth. Primary producers use solar energy to make organic compounds, including sugars, starches, and cellulose, from carbon dioxide and the hydrogen in water. Very nearly all non-photosynthesizers, including humans, depend on this process for their energy supplies because they consume primary producers either directly or indirectly.
- **Atmospheric gas regulation.** During photosynthesis, plants take in carbon dioxide and give off oxygen to make plant biomass. Plant growth results in a net withdrawal of carbon dioxide from the atmosphere and thus sequestration (locking up) of carbon in plant tissues. When the plants die, their decomposition releases : the carbon back to the atmosphere. Photosynthesis is also responsible for the existence of free oxygen in the atmosphere, the result of the labors over billions of years of photosynthetic bacteria. That oxygen, in turn, is the molecular pool from which ozone is formed in the upper atmosphere by the action of sunlight. Stratospheric ozone blocks most of the incoming ultraviolet-B radiation (UV-B is the most destructive frequency band in the ultraviolet range) and protects life on the surface from its damaging effects.

  Besides these functions, various organisms, especially bacteria, are critical links in the global geochemical cycles that keep essential elements like nitrogen, sulfur, and phosphorus in circulation and thus readily available. And in general, ecosystems modulate the gaseous composition of the atmosphere. That composition can change over time, as the action of photosynthetic bacteria illustrates, but ecosystems act to dampen those changes and keep them from being too abrupt.

- *Climate and water regulation.* Ecosystems and their components also have powerful effects on the hydrological cycle and on climate. Rooted plants pump a considerable volume of water from the soil and release it into the air via evapotranspiration. A single tree in a rainforest may move as much as 2.5 million gallons of water from the soil to the atmosphere in its 100-year lifetime. Besides thereby contributing water vapor (the most significant greenhouse gas) to the atmosphere, this action can also affect climate at a regional scale. In the Amazon basin, for example, water vapor from the Atlantic Ocean condenses and falls as rain, sinks into the forest soils, and is pumped back into the air by the trees and other vegetation. It then is carried inland by the winds, condenses again, and rains out farther to the west. This process is repeated several times. The moist rainforest climate is thus self-maintaining to an extent, and some scientists believe that reducing the forest cover below some critical threshold could disrupt the cycle and alter the region's climate.

  Not all the water that rains on vegetation is returned immediately to the air, of course. Trees and other plants break the force of falling rain and thereby allow it to soak into soil rather than running off. Much of the absorbed water then either percolates down into groundwater or is released gradually into springs, streams, and rivers. This process helps control flooding, stop erosion, and mitigate drought.

- *Soil formation and maintenance.* Good soil is rare. Only about 3% of the Earth's land area qualifies as high quality arable soil (Cleveland, 1994). Soil is also far more than just dirt; soils are actually complex ecosystems jam-packed with living creatures indispensable to soil function and persistence. A study of a square yard of Danish pasture, for example, revealed 40,000 earthworms, about the same number of insects and mites, and nearly 10 million roundworms. Forest soils have been found to contain millions of bacteria and tens of thousands of yeast cells and bits of fungus *per gram.* A gram of Iowa loam might contain billions of bacteria and hundreds of thousands of fungi, algae, and protozoa. These creatures don't merely inhabit the soil; they contribute to both soil fertility and waste decomposition. In fact, soil ecosystems are the main engines of decomposition of organic matter on the planet.

Soil microorganisms, by consuming dead matter, break it into its constituent elements (carbon, hydrogen, nitrogen, sulfur, and phosphorus) and free them for use again by plants.

Soil is partly formed by the weathering of rocks, but plants take part in the process by breaking rocks with their roots. Plants and animals also contribute carbon dioxide and organic acids that accelerate weathering. The decomposing microorganisms discussed above release carbon dioxide and water into the soil and leave humus, a complex colloid that consists of tiny particles of organic material of varying chemical composition. Humus helps support microorganisms and small animals like earthworms, as well as retain water and help give soil its texture.

As noted, plants not only help form soil but also retain it by anchoring it with their roots and protecting it from the impact of falling rain. Removal of soil plant cover generally speeds up soil loss to rates far in excess of rates of formation, which tend to be low. Depending on climate and other regional conditions, soil generally forms at a rate of 0.3 to 2.0 tons per hectare per year (a hectare is about 2.5 acres). Erosion rates, in contrast, range from about 16 tons per hectare per year in the United States to 40 or 50 tons per hectare per year in places like China and Zimbabwe (Cleveland, 1994).

- **Pest control and pollination.** Despite the occasional plague of locusts, most troublesome insects and diseases that threaten crops and domesticated animals are well controlled by natural ecosystems. Most of the pests are herbivorous (plant-eating) insects, and 99% of them are normally eaten by predator insects. Exceptions can occur when use of artificial pesticides wipes out the predators along with most of the target insects; since the pest populations tend to be much larger, they evolve resistance to the pesticide faster. Plant-eating insects are also good at developing resistance to pesticides because they have been engaged for millions of years in a kind of arms race with plants, which have developed poisons of their own to kill the bugs.

  As for pollination, despite the widespread use of domesticated honeybees, dozens of crop species (including alfalfa) depend on the fertilization services of natural ecosystems.

This brief look at the functions of biological diversity and natural capital should give some inkling of their value. We know from Chapter 2 that our capacity to substitute manufactured capital for natural capital is limited, and it should now be clearer why that is so: the goods and services provided by natural capital are fundamental, ubiquitous, indispensable, and vast in scale. We do not know how to make many substitutes, and when we do attempt it, they are often inferior in important ways to what they displace.

In view of natural capital's fundamental role in the life of the planet, including our lives, to say that it is difficult to put a dollar value on its functions is to say only that we have not yet succeeded where we must succeed. It is also to say that, in another sense, natural capital is beyond price. Yet, as will be seen in Chapter 4, the threats to the world's natural capital make the valuation effort far more than an academic exercise.

Chapter 4

Depletion and Valuation

WASHINGTON, D.C. — Virignia officials moved yesterday toward banning oyster harvests on the state's part of the Chesapeake Bay, where disease and overfishing have depleted the shellfish stock in the last three decades. ...The [state's] decision...was the latest sign of decline for the bay's once-bountiful fishery. Shad and sturgeon are nearly extinct. Rockfish catches are severely limited. ...Chesapeake Bay watermen routinely brought in more than 2 million bushels of oysters a season less than a decade ago. Last year, the bay yielded less than 175,000 bushels.

— D'Vera Cohn and John F. Harris
The Washington Post , August 25, 1993

SEATTLE — With a growing sense of shame, if not surren-
der, the Pacific Northwest is watching its wild salmon
dwindle into extinction. This year, the federal government
will impose the strictest fishing limits in the nation's history,
and it may go so far as to ban a salmon harvest in the
ocean waters north of Fort Bragg, California. ...[S]cientists
guess that 100 million salmon a year once emerged from
the rivers along the coasts of California, Oregon and Washington. Today, these fish are extinct in Southern California,
and the remainder of the region produces perhaps 15 million a year — most of them from hatcheries. Looked at
another way, 107 separate stocks of salmon have become
extinct in the Pacific states and 89 others are "at high risk
of extinction," according to Seattle-based Save Our Wild
Salmon, a coalition representing environmental and fishing
groups.

— John Balzar
The Los Angeles Times, April 3, 1994

Stories such as these are no longer uncommon. Moreover, since they are
stories in major daily newspapers with national readerships, and are about
large events close to home, they represent only the smallest sample of
episodes of natural capital depletion worldwide. In part because of the
poverty of our understanding of natural capital inventories, countless less
prominent examples go unnoticed; were it not for the Tellico Dam, few
would ever have known about the snail darter. But reported or not, the
depletion and degradation of natural capital are widening.

The most glamorous coverage of natural capital depletion is reserved
for such issues as global warming and the erosion of stratospheric ozone.
In the first case, it is virtually certain that concentrations of the so-called
greenhouse gases, especially carbon dioxide and methane, are increasing
due to human economic (industrial) activity. Concentrations of carbon
dioxide, for example, have increased by about one-quarter over pre-
industrial levels. There is also wide, though not unanimous, agreement
among scientists that such increases cannot go on indefinitely without
raising global average temperatures. Some argue that the climatological
records of the last century already show such a warming; others caution

that natural climate variability could account for the observed changes. In any case, the more critical debate now concerns the pattern, extent, timing, and consequences of future climate change, which is almost certain to be costly in economic and biological terms. This qualifies as depletion of natural capital because human activity appears to be exceeding the environment's capacity to assimilate wastes, and in so doing, creating conditions ripe for disruption of ecosystems and accelerated extinction of species.

The degradation of stratospheric ozone is an even clearer example of natural capital depletion. Stratospheric ozone blocks a majority of the sun's most damaging ultraviolet radiation (the medium-wavelength UV-B) from reaching the Earth's surface. UV-B has been linked to malignant and non-malignant skin cancer, weakening of the human immune system, diseases of the eye in humans and other animals, damage to the phytoplankton base of the marine food web as well as to fish larvae and juveniles, and damage to crops and natural vegetation, among other effects. First detected in the upper atmosphere over Antarctica, ozone depletion is spreading to the temperate regions and is likely to worsen over the next several years as atmospheric concentrations of already-emitted chlorofluorocarbon gases (CFCs), the chief culprits, continue to increase.

The 1987 Montreal Protocol and later efforts to phase out CFCs can be seen as a triumph of international cooperation in conserving a global resource. Put another way, these efforts reflect a global decision *to invest in natural capital* by ending production of the damaging agents and developing (at considerable expense) more benign substitutes, thus allowing the stratosphere's complement of ozone to begin repairing itself. This is truly a long-term investment, since it is likely to take a century or more for ozone concentrations to return to pre-CFC levels.

These are the dramas that make it into the headlines and the evening news shows. Many other measures of natural-capital depletion are less well-known but no less important. Some of them are briefly surveyed below.

Some Evidence of Natural Capital Depletion

Nonrenewable natural capital. In some ways the degradation of renewable natural capital is of greatest concern, since it contributes most directly to the reliable operation of the Earth's life-support machinery, an issue we will discuss in a moment. But first it is worth noting that nonrenewable resources are being depleted as well, as two examples illustrate. Oil production in the lower 48 states peaked around 1970 because the easily accessible oil had already been extracted and it became more expensive to drill for less-accessible oil than the prices received for the oil could justify (see Figure 4.1). Likewise, the quality of copper ore mined in the United States has been declining for decades (see Figure 4.2). (The "spike" occurred during the Depression, when only the richest mines remained in operation.) In both cases, the amount of energy required to extract a unit of resource continues to rise (Gever et al., 1991). This is of particular interest in the case of oil, because eventually it will cost more energy to get a barrel of oil out of the ground than is contained in the oil.

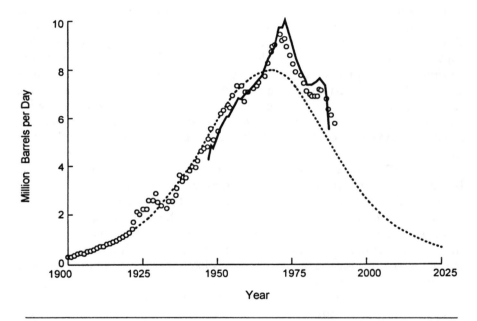

Figure 4.1   Annual rates of oil production in the lower forty-eight states. The dotted line represents the rate predicted by geologist M. King Hubbert in the 1950s, the solid line represents rates predicted by Kaufmann (1991), and the circles represent actual values. Source: Gever et al. (1991).

Land and agriculture. Food production has increased in every region of the developing world since 1970, mostly because of increases in yield rather than cropland expansion. On the other hand, these increases in yield have been accompanied by increases in the percentage of irrigated land, which has risen from 15% of total crop land in 1979-81 to 17% in

1989-91. Irrigation often salinizes soils and depletes groundwater supplies. At the same time, average annual fertilizer use went up from 81 kilograms per hectare of crop land in 1979-81 to 96 kilograms per hectare, an increase of 19%. The rate of growth in agricultural production has fallen from 3% per year during the 1960s to 2.2% per year in the 1980s, and global agricultural production actually fell in 1991 (World Resources Institute, 1992, 1994).

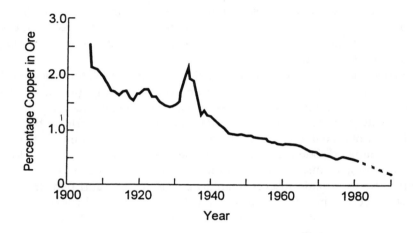

Figure 4.2   Ore quality of U.S. copper, 1906 to 1990. Source: adapted from Gever et al. (1991).

Most agricultural production in the world involves farming practices that are environmentally unsustainable. Efforts to encourage sustainable practices in the developed world are modest, while in developing countries, population growth and poverty retard the adoption of sustainable practices and encourage environmentally harmful expansion of agriculture (World Rescources Institute, 1992).

Unsustainable agricultural practices (short fallow periods, misuse of fertilizers, hillside cultivation, leaving soil exposed during fallow periods, compaction by heavy machinery, insufficient drainage), as well as overgrazing, deforestation, overexploitation for fuel wood, and industrialization (wastes, soil acidification from air pollution), have led to extensive soil degradation. Nearly 2 billion hectares of land worldwide have been eroded by wind and water, salinized, drained of nutrients, compacted and/or waterlogged. Agriculture accounts for 28% of the degradation, overgrazing about 34%, and deforestation 29%. About 1.2 billion hectares — 11% of the Earth's vegetated surface — have suffered moderate to extreme soil degradation since World War II (not including land degraded in the ancient past or land that is naturally barren). The 1990 United Nations-sponsored GLASOD study (Global Assessment of Soil Degradation) counted 750 million hectares as lightly degraded, 910 million hectares as moderately degraded, 300 million hectares as severely degraded, and 9 million hectares as extremely degraded. Extreme degradation is considered irreversible. The other grades require restoration efforts that vary downward in complexity and expense; many are beyond the means of most farmers in developing nations, and land that has lost its usefulness is often simply abandoned (World Resources Institute, 1992).

Agricultural pesticide contamination of groundwater sources has become incerasingly common in recent years, endangering local water supplies and polluting river and lake ecosystems in many nations. Pesticide contamination poisons bees and other pollinators, as well as birds that feed on contaminated seeds and predatory birds and mammals that feed on contaminated rodents. Heavy use of pesticides kills soil-dwelling insects and microorganisms, in effect sterilizing the soil. Modern pesticides are often toxic to aquatic insects, plankton, crustaceans, and fish. Ironically, insects, weeds, and plant diseases still destroy about a third of total crop production every year, roughly the same fraction lost before chemical pesticides were introduced some 50 years ago (World Resources Institute, 1994).

Forests and rangelands. Between 1980 and 1990, tropical forests shrank by 154 million hectares, a loss rate of about 0.8% per year, through conversion to other uses. During this period, deforestation rates in the Asia/Pacific region were 1.2% per year, 0.8% per year in the South America/Caribbean region, and 0.7% per year in Africa. Within these larger areas, the subregions of continental Southeast Asia and Central America/Mexico had rates about twice as high as the average for all the tropics. The rate is rising in continental Southeast Asia. According to studies by the United Nations Food and Agricultural Organization, Brazil and Indonesia apparently account for about 45% of global rainforest loss to date. These figures apply strictly to deforested regions; since the area of degraded and fragmented forest may be much larger than actual deforested area, habitat, and thus biodiversity losses, may be higher than deforestation numbers imply (World Resources Institute, 1994).

Besides the loss of habitat and biodiversity, deforestation adds greenhouse gases to the atmosphere. Deforestation in tropical countries accounted for losses of above-ground biomass estimated at 2.5 gigatons (billion tons) per year during the 1980s, the equivalent of 4.1 gigatons of

CO_2 (World Resources Institute, 1994). This is not an insubstantial amount; by comparison, U.S. total CO_2 emissions for 1990 were 5.0 gigatons (Energy Information Administration, 1993).

As for rangelands, almost all of the world's nearly 2 billion hectares of rangeland, except those in the Arctic, have been badly degraded by livestock grazing, introduction of exotic species, fuelwood harvesting, suppression of natural fires, and/or conversion to cropland or housing. Perhaps 35% of degradation is due to overgrazing (World Resources Institute, 1994).

Biodiversity. By various estimates, there are 3 million to 30 million species on Earth, perhaps 1.8 million of which have been identified. Although the scientific literature records the extinction of 58 mammal species and 115 bird species over the last 400 years, these are almost certainly underestimates, for three reasons: (1) many species that have never been identified are surely lost already; (2) the numbers do not include species lost in the last few decades, when the rate has been accelerating, since a species is not considered extinct until 50 years after its last sighting; and (3) most biodiversity is in the tropics, where habitat destruction is most extensive (and biota the least known) and has only happened recently. Various projections based on current trends in habitat destruction suggest that between 1 and 11% of the world's species *per decade* will be consigned to extinction during the next 20 years (World Resources Institute, 1994). E.O. Wilson argues that a "cautious" estimate puts the annual extinction rate at 27,000 (74 per day, 3 per hour) (Wilson 1992). The authors of one recent study of habitat fragmentation predicted that as many as 300,000 species now alive will disappear over the next 50 to 400 years; they term the loss an "extinction debt," since the die-off occurs several generations after the species' habitat is destroyed (McFarling, 1994; Tilman et al., 1994).

Water. Of all the world's vast supplies of water, only 2.5% is fresh, and nearly 70% of that is locked up in glaciers and permanent snow cover. The relevant figure concerning the fresh water available for human use is the rate at which the global hydrologic cycle renews fresh water; this figure is 41,000 cubic kilometers per year. However, most of that is lost to flood waters or is inaccessible. Probably no more than 14,000 cubic kilometers is actually available, and much of that should be left untapped to support natural ecosystems. Human per-capita use of fresh water has doubled since 1940 and so has the population, so that total demand for fresh water has increased by roughly a factor of four.

Stresses on available fresh water include contamination from agricultural runoff of surface water and groundwater, industrial contamination, injections of wastes into ground water, overpumping of slowly recharging aquifers, and use for cooling by nuclear and fossil power plants, which creates thermal pollution. But far and away the greatest stress is growing population; in 1990, 20 nations were classed as water-scarce, meaning that the per-capita availability of fresh water was below 1,000 liters per person per year. By 2025, up to 20 more nations will be classed as water-scarce. By 2025, the number of people living in water-scarce nations will fall between 817 million and 1.1 billion (Engelman and LeRoy, 1993).

Fisheries. The potential sustainable yield of fish from the world's oceans is estimated to be between 62 and 87 million metric tons per year. The former level was reached in 1984. Demand is increasing; the average annual catch of all types of fish from the oceans increased 29% during the 1980s, while the per-capita annual consumption of freshwater fish and seafood combined rose 17%. An estimated 1 billion people depend on fish as their sole source of protein.

The increase in the global marine catch during the 1980s stems from higher catch rates for species of lower value. Stocks of the most valuable species, such as cod and haddock, are declining. This is primarily due to fishing policies and practicies that encourage overexploitation. For example, governments subsidize their national fishing fleets by paying the fees due for fishing in foreign waters. Private owners also tend to invest heavily in boats and processing equipment when natural fluctuations drive fish stocks upward; when stocks subside, leaving the industry with too much expensive equipment, owners pressure governments for policies to protect jobs and investments. Overfishing and misguided policies have essentially destroyed some fisheries, such as the Peruvian anchovy industry and the New England groundfish industry (World Resources Institute, 1994).

The Green Devolution*

As suggested above and in the discussion in Chapter 2 on net primary productivity, agriculture is a potent factor in the alteration or degradation of natural capital. Human beings have been expanding agricultural production in several ways for thousands of years. Arable land and permanent pasture have been borrowed from natural ecosystems to the point that they now account for about 35% of the Earth's land surface. Perhaps half that area has been converted from wetlands, forests, and deserts in only

---

* Except where noted, this section is based largely on Cleveland, 1994, "Re-allocating work between human and natural capital," in Jansson et al., 1994, *Investing in Natural Capital: The Ecological Economics Approach to Sustainability.*

the last century. This vast expansion and the practice of multicropping (harvesting more than one annual crop from a plot of ground) are measures of the intensification of agriculture necessitated by growing human populations.

However, in the last 50 years or so, agriculture has intensified in a new dimension by increasing inputs of manufactured and nonrenewable natural capital in the form of machinery, water, fertilizers, and pesticides. These measures increase yields in the short term, but at a high cost in degradation of the natural capital base upon which agriculture ultimately depends.

The shift from traditional to industrial agriculture has involved two important kinds of substitutions. The first concerns the energy inputs necessary for cultivation, harvest, and pest control. At one time, these involved only human and animal labor. Now, the dominant energy sources are nonrenewable natural capital, e.g., fossil fuels. These are used directly to run machinery, pump water, and dry crops, and indirectly (along with phosphate rock) as feedstocks for agricultural chemicals and fertilizers.

The second substitution is in the knowledge and skills needed to farm. Success at traditional farming meant knowing about the local environment, soils, and climate, and knowing how to adapt farming practices and crop selections to the locale. Traditional farming knowledge encompassed a range of sophisticated strategies, including intercropping, agroforestry, terracing, and the use of biological pest controls. Farmers in the era of industrial agriculture must master a different repertoire: they do well if they are competent mechanics, can run crop spraying equipment, understand something of business, finance, and technology in general, and, increasingly, are computer literate. One result is that

> ...there has been a fundamental shift in the role of cultural capital away from the understanding and synchronization with local flows of renewable natural capital to the manipulation and control of large flows of nonrenewable natural capital that originate outside of the farm (Cleveland, 1994, p. 183).

The so-called green revolution was one highly touted outcome of this shift, and crop yields have indeed gone up significantly. But the reasons have little to do with higher crop efficiency and much to do with higher input subsidies. The photosynthetic efficiency with which plants can be made to convert sunlight into biomass and carbohydrate energy does not seem to have improved, on average. For example, natural ecosystems in North America convert solar energy to biomass at an efficiency of about 0.1%. The highest-yielding strains of corn and wheat, given generous inputs of water and fertilizer, can achieve as much as 0.5% efficiency. That's better than the natural rate, but most crops are much less efficient. Overall, crops are probably about as efficient as natural vegetation.

The higher crop yields have come primarily from the heavy increases in fossil fuel and other capital inputs (which have skyrocketed during this century; see Figure 4.3), and the extensive use of plant breeds designed to take full advantage of the inputs. In effect, we have created new plant strains and told them, "You just worry about producing edible biomass. We'll take care of the water, so you won't have to sink deep roots. We'll protect you from your natural enemies, so you won't have to spend energy producing your own pest-poisons. We'll kill the weeds, so you won't have to put energy into long stems and leaf cover to suppress them." To keep our end of the bargain — to create and maintain this pampered plant's entitlement paradise — we subsidize our crops with fossil fuel energy used to pump water for irrigation, operate farm machinery, and manufacture and apply fertilizers and pesticides.

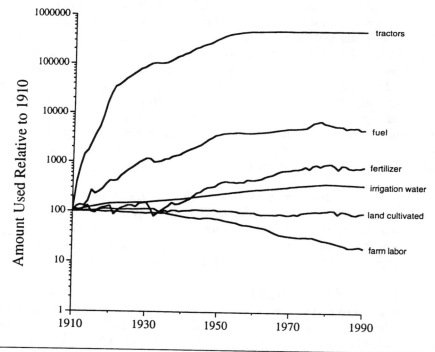

Figure 4.3   Inputs of various forms of natural and human capital to U.S. agriculture. Source: Cleveland (1994).

This sort of agriculture puts a severe strain on natural capital, both nonrenewable and renewable. Farm production is tightly bound to the ever-dwindling stocks of nonrenewable natural resources, especially fossil fuels. Even as global population and the need for higher agricultural outputs continue to rise, we become ever more dependent on inputs of resources that grow ever scarcer. The agricultural system is also quite sensitive to changes in the prices of the inputs, especially petroleum. Oil price shocks can buffet well-to-do farmers and devastate poor ones.

Besides draining stocks of nonrenewable natural capital, industrial agriculture also degrades renewable natural capital. For example, plants pure-bred as crops lose natural resistance to pests, and may no longer recycle nutrients as effectively. The use of pesticides destroys naturally occurring insect predators. Soils made productive by irrigation may become salinized, and thus irrecoverably less fertile, over time. Irrigation itself very often draws water out of the ground much faster than the natural recharge rate. Other industrial agriculture practices, such as mechanization, elimination of fallow periods, and harvesting of crop residues, accelerate soil erosion.

The off-farm impacts of industrial agriculture are illustrated by data for the United States, where 39 pesticides have been found in groundwater in 34 states. About 10% of U.S. community water systems show evidence of pesticide contamination, as do about 1% of all rural wells nationwide and up to 60% of private wells in Minnesota and Iowa. About 880 million metric tons of agricultural soils are carried off into reservoirs every year, reducing their flood-control capability and adding light-blocking sediments that harm submerged vegetation and thus the fish that need it for food and breeding habitat. U.S. soil erosion costs have been estimated at $10 billion annually. Artificially fertilized soils emit two to ten times as much nitrous oxide as unfertilized soils and pastures. Livestock and fertilizers account for very high percentages of ammonia emissions. Livestock and wet rice cultivation emit substantial amounts of methane, a potent greenhouse gas. Burning agricultural wastes causes emissions of CO, CO_2, and nitrogen oxides (World Resources Institute, 1992).

In the United States and elsewhere, our response to the loss of productivity caused by degradation of agricultural lands sometimes involves soil conservation programs and the like. More often, however, the response is simply to increase inputs. About half of current fertilizer inputs to U.S. farmland goes just to compensate for nutrients lost because of soil erosion. Another dramatic example concerns groundwater mining in the American Southwest. This semidesert has been made to bloom in a few areas by extensive and rapid withdrawal of water from the underlying aquifers.* When falling water tables in the Tucson, Arizona area doubled the cost of pumping irrigation water, an aqueduct was built to tap the Colorado River, 330 miles distant and 1,200 feet lower. By thus importing carrying capacity (water), Tucson vastly expanded its ecological footprint (see Chapter 2's discussion of carrying capacity) from the area of the local aquifer recharge zone to the area of the entire Colorado River watershed.

Responding in this way accelerates the rate of depletion of natural capital and ultimately raises the cost of food production. Bringing the Colorado River to Tucson drove up the energy cost of the area's water by a factor of five. Sustaining yields from eroded soils by increasing inputs of chemicals based on fossil fuels and phosphate rock raises the costs of those nonrenewable resources and thus the cost of food production, even as it continues the process of destroying soil fertility. The same kind of scenario plays out in pest control; the indiscriminate use of pesticides drives the evolution of resistant populations of pests, weakens or wipes out populations of natural predators, and leads in the end to the need for even heavier applications of pesticides and/or the development of deadlier poisons.

A Fearful Asymmetry: Accounting for Natural Capital

I look forward to the day when statisticians add up the national accounts to take account of the depreciation of the environment. When we learn to do this, we will discover that our gross national product has been deceiving us.

— Arthur Burns

Crucial to understanding human complicity in the loss of natural capital, when it seems so self-destructive, are the ways we value natural capital and measure its economic importance. We begin with gross national

---

* In some areas overlying the Ogallala aquifer, the world's largest, irrigation pumping is lowering the water table five feet per year. Officials of many of the states with access to the aquifer (Colorado, Kansas, Nebraska, New Mexico, Oklahoma, and Texas) made an open-eyed decision to use it up within 50 years at the most. (See Reisner, M., *Cadillac Desert: The American West and Its Disappearing Water*, Penguin Books, New York, 1986.)

product (GNP), which is probably the best-known economic statistic and possibly the most revered as well. Appeal is regularly made to the goal of boosting GNP in justifying all manner of economic policies. In the press, reassuring stories about rises in GNP, or disturbing ones about economic downturns (defined as slumps in GNP), appear constantly. GNP is monitored as an economic vital sign as closely as the pulse of a heart-attack victim.

This homage to GNP reflects the widespread, tacit assumption that GNP gauges, or at least correlates with, total human welfare; higher numbers must mean we're better off, lower ones worse. Economists know that this is, at best, an oversimplification. They know that when paramedics rush to an accident scene and carry a victim off to the hospital emergency room, the costs incurred make no net contribution to total welfare; instead, they're the costs of trying to restore the victim to his previous level of welfare (health). Likewise, the $300 cost of a catalytic converter on a new car goes toward keeping air quality from getting too much worse, not toward making it better — but it still adds to GNP. A famous example of GNP-growth perversity was the grounding of the Exxon *Valdez* in Alaska's Prince William Sound in March 1989. Total costs for cleanup, compensation, and litigation have run into the billions — all of it helping to boost GNP. No doubt a few people did very well out of the accident, but net total welfare would surely have been higher if the ship had missed the reef.

For such reasons, many economists appreciate that GNP is not a particularly good index of total economic welfare. Its recognized short-comings include

> ...the failure to allow for externalities (e.g., the failure to include pollution, congestion and so on, as negative items); the difficulties involved in evaluating home-produced output or subsistence output, which may be very important in poorer countries; the common failure to value the output of most public services in a welfare-oriented manner instead of in a cost-of-input manner; the failure to include the output of many public facilities and the failure to allow for many other non-market activities, of which housewives' services are one of the best-known examples; the failure to allow for differences in leisure or working conditions; and, above all, the failure to reflect the degree of equality in the distribution of income (Beckermann, 1991, p. 487).

Not explicitly mentioned in this list is the failure to account for depletion of natural capital, in part because of the reasons discussed in Section I concerning neoclassical economics' demotion of natural resources to the status of a rather trivial factor of production. At the global level, however, the state of the stocks of natural capital is fundamental to long-term economic welfare. This is frequently true at the national level as well. Although nations like Japan and Switzerland maintain very high standards of living without abundant resource stocks of their own, they do so simply by importing natural resources in great quantities from resource-rich nations. Nations that depend heavily on the export of natural resources for income, as do a number of developing nations, can easily deceive themselves about their long-term prospects if their method of national income accounting does not capture the real value of natural capital.

GNP and its cousin GDP (gross domestic product) are parts of a general system of economic accounting called the System of National Accounts (SNA) that misvalues, undervalues, or simply ignores natural capital (Repetto, 1992). The SNA has been around for 50 years, and although efforts are being made to alter it, it still reflects the economic concerns — which did not include resource scarcity — of the immediate postwar era when it was devised. What it does clearly reflect is the tendency of most economists since the classical era to treat natural capital wholly differently than manufactured capital. Manufactured capital is depreciated in recognition that its value declines each year as it wears out or is used up, but the use or loss of natural capital is not counted as a charge against current income. Natural capital is treated in these accounts as if it had no value.

Natural capital, as we have already seen, is crucial to economic development, nowhere more so than in resource-dependent developing countries. So it is not necessarily the use of natural capital that is troublesome, but the failure to understand what its use means, to account properly for it, and to invest in it:

> The revenues derived from resource extraction can finance productive investments in industrial capacity, infrastructure and education. A reasonable accounting representation of the process, however, should recognize that one kind of asset has been exchanged for another. Should a farmer cut and sell the timber in his woods to raise money for a new barn, his private accounts would reflect the acquisition of a new income-producing asset, the barn, and the loss of an old one, the woodlot. He thinks himself better off because the barn is worth more to him than the timber. In the national accounts, however, income and investment rise as the barn is built, and income also rises as the wood is cut. Nowhere is the loss of a valuable asset reflected. Even worse, if the farmer used the proceeds from his

timber sale to finance a winter vacation, he would be poorer
on his return and unable to afford the barn. But national income
would still register a gain (Repetto, 1992, p. 96).

The failure of the SNA to capture the value of natural capital has
consequences that extend beyond entries in a ledger. Decisions about
when, how, and how fast to use natural capital are ordinarily evaluated
according to several criteria, including the prevailing rate of interest (see
sidebar, "When 'discounted' doesn't mean a bargain"). The decision framework takes for granted that: (1) all economic values of the natural capital
are known and reflected in the prices the resources bring on the open
market; (2) markets are undistorted by subsidies, externalities, and so on;
and (3) the proceeds from the extraction and sale of the resources will
actually be reinvested in other productive capital, not just squandered
(Barbier, 1994). None of these is always true, and it is safe to say that
the first has probably never yet been true. Timber extracted from a forest,
for example, is typically priced simply as a commodity; no prices (and
no values) are attached to the forest as habitat, provider of home and
livelihood for indigenous people, protector of watersheds, and maintainer
of microclimates or biodiversity storehouse. The result is that natural capital
may be casually or wantonly used up or destroyed because

> ...we do not know or bother to take into account the potential
> benefits it yields. As a result, decisions will always be biased
> toward environmental degradation because the underlying
> assumption is that the foregone benefits provided by natural
> capital...are necessarily negligible (Barbier, 1994, p. 296).

These issues become acutely important for developing countries that
depend heavily on resource exportation for income. As the tables show,
export income for a number of developing nations is very narrowly based
on one or two primary commodities (see Tables 4.1 and 4.2). In many
cases, the primary commodities have accounted for all or most export
income over a period of more than 20 years, suggesting that very little
reinvestment is going toward broadening the economic base. And yet
these are the very places where it is most imperative to manage the natural
resource base carefully in order to ensure long-term earnings from exports
and promote the transition to a state of lower resource-dependency.
Additional pressure comes from the fact that most of these countries'
external debts, and the interest on their debts, have risen sharply relative
to GNP and export income. They will have to manage their natural capital

Table 4.1   Low Income Economies with High Export Concentration in Primary Commodities [a]

Contribution of Primary Commodities to Total Exports [b]	Export Share in 1980/81	Export Share in 1965	Main Export Commodities [b] 1		2	
over 90%				%		%
Uganda ($280)	100	100	100	Coffee	95.6	Tea	[c]0.3
Eq. Guinea ($410)	100	91	NA	Cocoa	34.4	Coffee	3.1
SAO Tome & pr. ($490)	99	100	NA	Cocoa	95.5	Copra	1.8
Ethiopia ($120)	99	99	99	Coffee	53.8	Hides	15.3
Rwanda ($320)	99	99	100	Coffee	75.5	Tea	[c]10.8
Yemen PDR ($430)	99	NA	94	NA		NA	
Zambia ($290)	98	99	100	Copper	83.0	Cobalt	5.4
Burkina Faso ($210)	98	85	95	Cotton	32.6	Livestock	[e]26.8
Nigeria ($290)	98	99	97	Petroleum	87.3	Cocoa	4.7
Liberia ($450) [d]	98	98	97	Iron Ore	63.4	Rubber	[e]16.1
Ghana ($400)	97	98	98	Cocoa	51.1	Gold	20.3
Mauritania ($480)	97	99	99	Fish	65.8	Iron Ore	33.3
Niger ($300)	96	98	95	NA		NA	
Somalia ($170)	95	99	86	Meat	39.7	Banana	34.5
Zaire ($170)	93	94	92	Copper	35.8	Coffee	[e]11.2
Sudan ($480)	93	99	99	Cotton	30.3	Livestock	24.4
Togo ($370)	92	85	97	Phosphate	36.2	Cotton	12.6
Comoros ($440)	[c]92	[e]86	NA	Cloves	41.7	Vanilla	[c]33.3
Lao PDR ($180)	90	[e]100	NA	Timber	51.7	Electricity	19.0
over 80%							
Chad ($160)	[c]89	[e]96	97	Cotton	69.4	Hides/Skins	[e]3.8
Myanmar ($210) [c]	89	[d]81	NA	Rice	32.7	Teak	[e]32.2
Guinea-Bissau ($190)	[c]87	[d]71	NA	Cashewnut	73.3	Groundnut	[c]6.7
Guyana ($420)	[c]87	NA	NA	NA		NA	
Madagascar ($190)	84	92	94	Coffee	26.6	Cloves	5.7
Malawi ($170)	83	93	99	Tobacco	62.8	Tea	10.3
Burundi ($240)	83	96	95	Coffee	82.6	Tea	5.0
Kenya ($370)	83	88	94	Coffee	26.2	Tea	[c]21.9
Tanzania ($160)	81	86	87	Coffee	31.4	Cotton	12.7
over 70%							
Maldives ($410)	[c]77	[e]70	NA	Fish	[c]57.1		
Benin ($390)	74	96	95	Cotton	13.4	Fuel	9.4
Indonesia ($440)	71	96	96	Petroleum	40.0	Rubber	5.0
Mali ($230)	70	83	97	Cotton	36.9	Livestock	29.0
over 60%							
C.A.R. ($380)	60	74	46	Diamonds	40.9	Coffee	18.9
over 50%							
Sri Lanka ($420)	57	79	99	Tea	25.9	Rubber	[c]7.0

Notes: a/ Low-income economies are those with per capita incomes of $545 or less in 1988. U.S. dollar figure at country listed indicates GNP per capita in 1988. b/ Contributions to the value of total merchandise exported 1988, unless indicated. c/ 1987 value. d/ 1981–83 average value. e/ 1984 value.

Source: Based on various editions of the following World Bank documents: *World Development Report; Trends in Developing Countries; Commodity Trade and Price Trends; African Economic and Financial Data.*

Table 4.2. Lower Middle-Income Economies with High Export Concentration in Primary Commodities [a/]

Contribution of Primary Commodities to Total Exports [b/]	Export Share in 1980/81	Export Share in 1965	Main Export Commodities [b/]			
			1		2	
over 90%				%		%
Bolivia ($570)	97	100	95	Gas	40.1	Tin	13.9
Papua N.G. ($810)	95	100	90	Gold	37.7	Copper	28.9
Ecuador ($1120)	93	93	98	Petroleum	44.8	Fish/Shrimp	19.0
over 80%						
Yemen A.R. ($640)	89	49	100	Oil	93.7		
Honduras ($860)	89	89	96	Bananas	39.0	Coffee	21.0
Congo ($910)	89	94	37	Oil	71.6	Timber	15.6
Cote d'Ivoire ($770)	88	90	95	Cocoa	25.7	Coffee	13.1
Cameroon ($1010)	88	97	94	Petroleum	48.9	Coffee	12.2
Paraguay ($1180)	88	NA	92	Cotton	10.3	Timber	2.5
Chile ($1510)	85	90	96	Copper	48.4	Agriculture	13.2
over 70%						
Panama ($2120)	79	91	98	Petroleum	31.7	Banana	[d/]17.9
Peru ($1300)	78	83	99	Copper	12.9	Zinc	8.8
Senegal ($650)	75	81	97	Fish	26.9	Groundnut	14.8
Columbia ($1180)	75	72	93	Coffee	30.2	Oil	17.0
Syria ($1680)	75	NA	90				
Egypt ($660)	74	92	80	Oil	64.4	Cotton	6.5
Dominican Rep. ($720)	74	81	98	Nickel	31.2	Sugar	20.5
El Salvador ($940)	71	63	83	Coffee	60.6	Fish	[c/]3.5
over 60%						
Guatemala ($900)	62	71	86	Coffee	34.8	Banana	7.6
Zimbabwe ($650)	60	63	85	Tobacco	21.5	Gold	13.1
Costa Rica ($1690)	60	68	84	Coffee	30.4	Banana	18.8
over 50%						
Malaysia ($1940)	55	80	94	Rubber	9.8	Palm Oil	8.4
Jordan ($1500)	53	57	81	Minerals	38.6	Food	10.2
Brazil ($2160)	52	59	92	Soya	9.4	Coffee	6.8
Morocco ($830)	50	72	95	Ph. Acid	16.8	Phosphate	13.3

Notes: a/ Low-income economies are those with per capita incomes of $2160 or less in 1988. U.S. dollar figure at country listed indicates GNP per capita in 1988. b/ Contributions to the value of total merchandise exported 1988, unless indicated. c/ 1987 value. d/ 1984 value.

Source: Based on various editions of the following World Bank documents: *World Development Report; Trends in Developing Countries; Commodity Trade and Price Trends; African Economic and Financial Data.*

base very carefully to pay off those debts and induce further economic investment (Barbier, 1994).

Costa Rica provides a case study of the way in which failure to account properly for natural capital depletion can wreak ecological destruction and hamstring a nation's economic flexibility (Repetto, 1992). Since about 1970, Costa Ricans have burned off or otherwise cleared at least 30% of the nation's forest cover. Converted to fields and pastures and exposed to heavy rainfall, the land suffered savage erosion, losing an estimated 2.2 billion tons of soil between 1970 and 1989. In parallel, coastal fishing grounds were heavily damaged by water pollution and overharvesting. The corresponding economic losses, as estimated in a study by the Costa Rican Tropical Science Center and the World Resources Institute, were severe:

> The year 1989 saw the destruction of 3.2 million cubic meters of commercial timber worth more than $400 million. This amount, $69 for each person in Costa Rica, exceeded payments on the foreign debt by 36 percent. Erosion from farmland and pastures washed away nutrients worth 17 percent of the value of the annual crops and 14 percent of the value of livestock products. The deterioration of stocks in the main fishing ground was so severe that fishermen's earnings fell beneath the level of welfare payments to the destitute. ...From 1970 to 1989, the accumulated depreciation in the value of [Costa Rica's] forests, soils and fisheries exceeded $4.1 billion in 1984 prices — more than the average value of one year's GDP (Repetto, 1992, pp. 97-8).

Note that this estimate does not include the value of the loss of unpriced forest goods and services, most fish species in the fisheries or several kinds of soil characteristics. Even so, Costa Rica clearly did far less well economically than the conventional accounting system, which ignored these losses, suggested. The net rate of capital formation (how rapidly natural and manufactured capital are created), for example, was overstated by more than 70% for 1989.

Costa Rica is not the only nation that has clearly suffered from faulty accounting of natural capital depletion. Indonesia is another; studies similar to those performed in Costa Rica revealed that while Indonesia's GDP rose at a nominal rate of 7.7% per year from 1970 through 1984, corrections for depletion of natural capital slashed the rate in half, to 3.9% per year — and even that was an overstatement, since the studies accounted for depletion of oil, timber, and soils only, and ignored exploitation of several other resources (Repetto, 1993).

These examples hint at some of the major defects of the current system of national accounts (Prince and Gordon, 1994). The national accounts fail to accurately describe changes in the quality and quantity of natural capital stocks, in contrast to the way stocks of manufactured capital are treated. Factories and machinery are counted as productive capital, and when they wear out or suffer accidental damage it is labeled capital consumption and the value of the loss is subtracted from GNP. Not so for natural capital; the wear and tear on forests, soil, air, and water caused by their exploitation is not subtracted from their value.

Further, the accounts that summarize input and output do not explicitly show the value of the services provided by ecosystems in absorbing and "processing" wastes, nor the value of defensive expenditures against pollution, such as pollution abatement equipment, home air filters, medicines and medical treatments, and so on, which are categorized with other investment items. The accounts also do not reflect the income value of nonmarket services such as recreation on public lands, aesthetic benefits, the potential benefits of biological diversity, or so-called nonuse benefits, which are those derived from simply knowing, for example, that an endangered species is surviving even if one never sees it or that wilderness areas still exist even if one never visits them.

Three kinds of changes have been proposed to address the shortcomings of the national accounts (Prince and Gordon, 1994). The first, called expanding the asset boundary, is to broaden the set of goods we treat as national assets (capital) to include the now-uncounted environmental and natural resources. This would involve estimating values for the depletion and degradation of forests, mineral reserves, and agricultural lands. The second change, called expanding the production boundary, would be to redefine national income to include the value of recreation, aesthetics, biodiversity, and nonuse benefits. Finally, it has been suggested that items within the production boundary should be reorganized so that the system recognizes the contribution to the production process of environmental services such as waste absorption. It might also involve reclassifying "defensive" spending, such as that for pollution control and abatement, as so-called intermediate expenditures that would be subtracted from GNP.

Although the case studies of particularly resource-dependent countries described above illustrate the direst consequences of failing to account properly for natural capital, no national accounting system anywhere is yet fully adequate to the task. The ubiquity of this problem and the acknowledged shortcomings of the national accounts have prompted efforts by the United Nations and by individual governments to devise accounting systems that give a better grip on the value of natural capital. The U.S. Department of Commerce's Bureau of Economic Analysis and the UN Statistical Office have already begun research to incorporate more information on natural resources and the environment into national accounts. They are trying to identify the costs of pollution abatement and control and the monetary value of environmental degradation, but are not yet working on redefining the production boundary to include now-excluded services of natural capital.

Thus far we have focused only on the failure of GDP and GNP to properly credit the fundamental contributions of natural capital to human welfare. In passing it is worth noting that these measures fall short in other ways as well, when they are taken as proxies for well-being. For example, since 1990 the United Nations Human Development Program has published an annual report ranking the world's nations on a scale that factors in not only per-capita wealth but also life expectancy, quality of nutrition, mean years of education, access to water and sanitation, the evenness of the distribution of wealth, and other measures (including, at one time, a "freedom index" reflecting the extent of personal liberties) (Lewis, 1993). By the Human Development Index (HDI), the high per-capita wealth of Saudi Arabia, which puts it 31st on the GNP-per-capita list, cannot make up for the great disparities between the rich and the poor; it ranks only 84th on the HDI list. The United States, which provides more years of schooling for its citizens than any country (12.3 years), ranks 6th on the HDI, even though in terms of per-capita wealth alone it ranks 10th. Canada ranks 2nd on the HDI but 11th in per-capita wealth.

Another attempt to devise a more comprehensive and honest measure of economic well-being is the Index of Sustainable Economic Welfare (ISEW) (Daly and Cobb, 1989). Very briefly, ISEW considers and melds the following factors:

Income distribution. Typically, a poor household will benefit more by an increase in income of X dollars than will a rich one. Put another way, X dollars given as an increase in income will benefit the poor household more than the loss of the same amount would hurt a rich household. Wide gaps between the richest and poorest groups in a society tend to reflect greater concentrations of wealth in the hands of a few, i.e., greater inequities in distribution. ISEW incorporates a measure of the difference between the incomes of the richest one-fifth of the population and the remaining lower-income fifths.

Net capital growth. If population grows, economic welfare will decline unless the stock of capital (from which income flows) increases to keep pace. ISEW measures total net capital growth by adding increases in manufactured capital, less the amount required to maintain the same per-capita level. Land (a form of natural capital) is not counted because of a data problem; the measure of increase is rising land prices, which really only reflect changes in demand, not changes in the capital stock (amount of land), which is forever fixed. Neither is human capital counted, although the authors believe it should be in principle, because the things usually used to assess the state of human capital stocks — years of education, extent of medical care, etc.— actually have rather ambiguous relationships to human capital. For example:

> ...the correlation between earned income and education appears to be very weak. Jacob Mincer, one of the leading analysts of investments in human capital, has shown that, among white, male, nonfarm workers, only 7 percent of the variation in earned income is accounted for by differences in their levels of education. ...In other words, 93 percent of the variation is due to other factors, ranging from luck and personal connections to ambition, native ability, and skills learned on the job (Daly and Cobb 1994, p. 404).

Natural resource depletion and environmental damage. ISEW attempts to account for the "depreciation" of natural capital by subtracting an estimate of the income lost to future generations by the current depletion of exhaustible resources such as fossil fuels and other minerals, as well as by the loss of biological resources. It also attempts to account for environmental damage in the form of possible changes in climate, air and water pollution, and noise pollution.

Unpaid household labor. Much work is done in households, mostly by women, that is nominally unpaid and yet is of enormous value. There are great conceptual difficulties in assessing this value. For example, in surveys people variously classify child care as work, as leisure, and as both. Further, the most straightforward way of arriving at a dollar value is to count the total hours spent in household labor and multiply that number by the typical wage for domestic help; however, this seriously undervalues the managerial function of the person who directs the housework of others. Yet the contribution of unpaid household labor to economic welfare is much too significant to ignore. ISEW uses the domestic-help method just mentioned.

In addition to these measures, ISEW also adds the value of expenditures on good streets and highways and on public health and education, and subtracts such items as defensive expenditures on health and education (trauma care, remedial reading programs, etc.), spending on national advertising (because it tends to promote demand, while local advertising is more informational in content), and the costs of urbanization, commuting, and auto accidents, among other things. The end result is expressed as a dollar value that can be compared with per-capital GNP, the traditional measure of economic well-being. As Figure 4.4 shows, making these adjustments reveals a considerably lower average level of welfare in the United States. Whereas per-capita GNP rose to over $7,200 in 1986, per-capita ISEW was less than half that, and actually peaked in 1976.

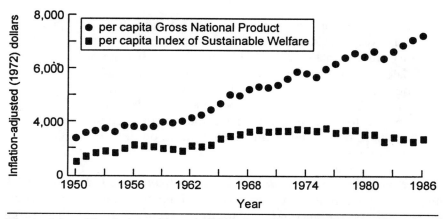

Figure 4.4   The index of sustainable economic welfare versus per-capita GNP in the United States. Adapted from Daly and Cobb (1989).

The Monetary Valuation of Natural Capital

In order to account for natural capital, it must be priced. This turns out to be difficult but possible in many cases; for example, it is fairly straightforward when an element of natural capital is bought and sold on the open market. In those cases, at least the marketed function of the natural capital can be assigned a value. In the case of natural capital functions for which there are no markets, values can often be imputed or estimated by various methods that yield "shadow" prices.

We will discuss some of those methods in a moment, but first let us note that many people believe that pricing the natural world cannot be done morally. What price should be assigned to, say, an obscure, nondescript butterfly species threatened with extinction by the conversion of its last remaining acres of habitat into a golf course? An unconstrained market would say zero, since the butterfly has no economic value. Nevertheless, many people would feel a strong sense of loss if it should

When "Discounted" Doesn't Mean a Bargain

> Discounting can easily become a pseudoscientific way of making the ethical judgment that the future is not worth anything.
> — Herman Daly,
> *Steady-State Economics*

Because of the existence of interest, a small sum of money invested today can grow to be a much larger sum in a few years (the one known exception being college-education funds, which mysteriously never seem able to grow large enough). Put a dollar into an investment earning 10% per year and it will be worth $1.10 a year from now, $1.21 a year after that, and so on. Turning the idea around gives us this: because of interest, something that may be very valuable (in monetary terms) in the far future is not worth very much right now (i.e., its "discounted present value" is low). A dollar 20 years from now is "worth" only 11 cents today, because it would earn enough interest at our 10% rate to equal a dollar in 20 years.

This is merely the economic expression of the common belief that, since life is uncertain and the future unpredictable, a benefit or a loss is more important if it happens now than later. That is, the present value of a benefit (or loss) is greater than its future value. This belief is the basis for *discounting* the future. The rate used for finding the present value of some future benefit or cost is called the discount rate. Discount rates ordinarily are set more or less equal to prevailing rates of interest.

Discount rates can be enormously important in the management of natural capital, especially renewable natural capital. To see why, we must touch on the work of Harold Hotelling, a Stanford University economist whose interests in the 1930s lay in the problem of the economically most efficient way to extract a natural resource (Clark, 1989). Hotelling devised a method to calculate the best choice among the three main possibilities. Suppose a person owns a stand of immature trees. As they grow, they increase in value. The forest owner can (1) cut the trees down and sell them now; (2) hold on to them, letting them grow in both size and value, and/or in the hope that timber prices will rise and thereby increase profits from harvesting later; or (3) try to figure out the optimal rate of steady harvesting, which would spread out profits over time.

Our forest owner would consider several factors in making this decision. What is the price of timber? How much does it cost to cut the trees down and get them to market? How long will the trees last at a given rate of harvest? What is the going rate of real interest (the nominal rate of interest minus the rate of inflation)?

This last question is key, for it turns out that if the real rate of interest is higher than the rate at which the trees increase in value, then the only *eco-*

nomically sensible course is to cut down all the trees right away, sell them and put the money into an investment yielding the higher rate. Taken alone, Hotelling's model argues that natural capital should be converted to manufactured, or reproducible, capital if the latter will earn a higher return.

It should be obvious why this can be a problem for natural capital. Interest rates create a kind of temporal chauvinism, expressed in discounting. By embodying in economic practice our belief that the future is less valuable than the present, they tend to create a bias against conservation and undermine efforts toward sustainability. This is especially true of high discount rates, since a great many valuable or potentially valuable species do not grow or expand very fast. For example, teak and other hardwood trees grow so slowly that they are unlikely to be replanted after cutting unless interest rates are low. On the other hand, eucalyptus trees grow fast enough that they may be profitable even when interest rates are high (Costanza et al., 1994). Discounting thus can lead to overexploitation and depletion of natural capital without reinvestment, and can alter ecosystems by biasing investment toward faster-growing species.

Another problem is that discounting the value of a development project's future environmental damage makes the value of the harm seem low compared to the present costs of avoiding it. This is one reason why it is difficult to drum up much real enthusiasm among the nations of the world for action to avert possible global climate change.

Using market rates of interest as a guide to setting discount rates doesn't necessarily distribute the burdens and benefits of investment decisions fairly between the present and the future (Stiglitz, 1986; Daly and Cobb, 1989). The supposed basis of markets is the self-interested actions of mortal individuals. The notion of discounted present value represents

> ...the value to present people derived from contemplating the welfare of future people. It does not reflect the welfare of future people themselves, or even our estimate of their welfare. Rather it reflects how much we care about future people compared to ourselves (Daly and Cobb, 1989, p. 154).

However, communities outlive their individual members, and what may make sense for individuals seeking to maximize short-term present value — e.g., over-harvesting a fish species to extinction and putting the money in securities markets — may threaten the longer-range interests of communities.

The selection of discount rates is often a contentious and politicized process. There is both reason and room for maneuver in selecting discount rates, and concerns such as those discussed previously have prompted some environmentalists to urge use of low rates with projects entailing large environmental impacts.

However, economists are not unanimous on the effect of interest and discount rates on natural capital management, even when they are sympathetic to the need to conserve natural capital. Some argue, for example, that although high interest rates tend to encourage depletion of resources now and thus shift ecological costs to the future, high rates also discourage investment in general, since it costs too much to borrow money and few projects will earn a high enough return to pay off loans or compete with leaving the money in the bank. Since natural capital is necessary for investment (it takes resources to build a dam or a factory or a housing development), when investment is low, so is the demand for natural capital. Thus, "[e]xactly how the choice of discount rate impacts on the overall profile of natural resource and environment use in any country is...ambiguous" (Pearce and Turner, 1990, p. 224). These analysts argue that it would be better to build sustainability into economic decisions by setting an *a priori* requirement that the total stock of natural capital be left constant, regardless of the other benefits and costs.

disappear. In the United States, laws such as the Endangered Species Act reflect the presumption that the butterfly has value apart from economic considerations. Ironically, these can have the effect of creating a facsimile of economic value. The butterfly species' price might be said to lie within a range bounded at the lower end by the legal costs of filing an environmental impact statement and perhaps litigating the matter, and at the upper end by the foregone value of the golf course, should the decision to protect the species block any development at all.

Take another example: What is the value of the Pacific wild salmon whose once-vast stocks have declined so sharply in recent years, primarily because of hydropower development and degradation of spawning streams? Commercially, wild salmon were once of considerable value. Now they are nearly worthless, even though they have become scarcer. Salmon is plentiful in the stores and not much more expensive than hamburger — thanks to vastly increased fish-farming. The fish are not wild salmon, and the high-input industrial fish farming that produces them is not sustainable, but the market and consumers (acting as "rational economic persons") do not care and thus create no market-based pressure for the preservation of wild salmon stocks.

Markets clearly ignore certain important values. For example, in neither case, butterfly nor salmon, is any value imputed to the species' role as a component of an ecosystem. For many species, this value might be substantial. Ask a pilot flying an airplane which pieces — wings, tail, fuselage, or engine — he or she would be willing to part with, and the answer will be "none of the above, for any amount of money." Broadly speaking, they must all be present and working as an ensemble for any of them to be of any value at all. The value of the pieces as parts of a system is vastly greater — infinitely greater, if losing one means the difference between flight and doom — than the sum of their individual market values. So it often is with the elements of natural capital that make up ecosystems, although such values may be difficult or impossible to estimate at present.

Those who argue from a moral point of view also say that such considerations completely miss the point. Species extinction, many might say, is an assault on the integrity of the ecosphere that transcends dollars and cents. Species have a right to exist, period, and attempting to assign values to them is meaningless.

In spite of the repugnance with which many people view the assignment of dollar values to bits and pieces of natural capital, it is a useful, if incomplete, yardstick. It permits reasonable comparisons to be made among different benefits and costs, or gains and losses, and it yields a quantified measure of peoples' preferences (Pearce and Turner, 1990). And while nonmaterial values are important to many practitioners of ecological economics, it remains, after all, an economic approach. Although the monetary yardstick cannot capture all the values of natural capital, and there are other problems with it as well, in this section we will set them aside in order to see what use can be made of it.

As yet there seems to be no universal agreement on the ways to classify or label the values of natural capital, and the various schemes sometimes overlap. Probably the simplest scheme defines three categories of value: use, option, and existence (Pearce and Turner, 1990). *Use value* is that which derives from the direct use or consumption of natural capital products. Sometimes use takes place within the economy, in which case the values for at least some of the functions of the natural capital are captured by the market (and usually reflected in the national accounts). For example, a tree can be sold for lumber and the sale price will represent the market value of the tree as a building material. (This assumes, of course, that the harvesting was not subsidized in some way, which is not uncommon and which would mean that the market price understated the tree's real value, even as a commodity. And its value as a nesting place for a spotted owl will not be reflected in the price either, but that is not considered a use value.) In a great many other cases, natural capital products are consumed outside the market (e.g., wildlife harvested by indigenous peoples) and are not reflected in the national accounts. These two types of use value (market-captured and market-ignored) have also been called productive use value and consumptive use value, respectively (De Groot, 1994).

Option value generally refers to the value of potential uses of natural capital, in contrast to its present use value. Included in this category are the vicarious benefits derived from knowing that someone else might be able to secure a use value from natural capital. Also included is bequest value, which derives from preserving natural capital for the use of future generations.

The third category, *existence value*, is defined more or less as those values not covered by the other categories, i.e., as the value of natural capital unrelated to any current or potential uses. Existence value covers the sense of sympathy or fellow-feeling people have for other species even when they do not use them or even have any contact with them except perhaps via photographs or television programs. Bottom line, existence value rests on human altruism or sense of stewardship for the Earth's creatures (Pearce and Turner, 1990).

To these three could also be added *conservation value*, which is the value of natural capital functions such as climate regulation, topsoil formation, and maintenance of the chemical composition of atmosphere (De Groot, 1994). These are environmental services that we rely on and which form a kind of infrastructure for the economy, although they do not provide (apart from making life possible!) direct economic benefits.

How does one put numbers to the nonmarketed values of natural capital? There are several methods, though they have their limitations and are not totally free of controversy.

One such method, called hedonic pricing, depends on the analysis of the various attributes of a piece of land or property that make it valuable (Prince and Gordon, 1994). A property such as a house is generally valuable to the extent that it is in good shape, spacious, near stores and schools, and so on. Its value will be higher if it can boast more of the things people generally want in a house.

Now suppose, for example, that we want to come up with a measure of the value of air quality in a particular area. A house located just downwind of an aluminum smelter or a coal-fired power plant will probably be worth less than an otherwise identical house located where the air is clean. Of course, no two houses are identical, but by examining large numbers of houses it is possible statistically to hold all variables constant except air quality and, by seeing how average prices of comparable houses differ in areas of clean and polluted air, arrive at a measure of what it is worth to homebuyers. Studies of this sort suggest that a 1% increase in the levels of various air pollutants can drive down property values by as much as 0.5% (Pearce and Turner, 1990).

Another common technique is called the contingent valuation method (CVM) (Pearce and Turner, 1990). CVM simply asks people what they would be *willing to pay* to receive a benefit or to prevent a loss (WTP), or what they would be *willing to accept* to forego a benefit or tolerate a loss (WTA). CVM could be used to estimate an existence value for the vanishing Pacific salmon, for example, by asking a representative sample of people living in the Northwest what they would be willing to pay in the form of higher electricity rates and consumer prices, which might result from modifying hydroelectric facilities to allow for easier transit by the fish during spawning season and for reducing the industrial pollution that damages spawning grounds.

One problem with this approach is that residents of the Northwest might be willing to pay a lot to protect "their" Pacific salmon stocks, but much less to protect other equally deserving salmon species elsewhere. Another frequent criticism of WTP and WTA studies is that they depend on what people say they would do, rather than what they actually would do. In one study, for example, Montana fishermen were asked if they would be willing to pay to protect two rivers; 6.6% said yes. But when a subgroup of the original respondents was asked for donations, only 1.1% sent money (Passell, 1993).

It is also possible that asking such willingness-to-pay questions in a vacuum might bias the results. That is, the amount of money someone might be willing to pay to save the Pacific salmon could easily be much higher if that is all he or she is asked to take on than if the list also includes paying to repair the ozone, reduce carbon dioxide emissions, stop topsoil erosion, reduce the use of petroleum-based pesticides, prevent overharvesting of old-growth forests, and address all the other environmental problems caused by too much throughput. Yet another complication is that a rich person could pay more than a poor one, even though their desires to preserve salmon might be emotionally equivalent. Does that mean the value of salmon depends on the wealth of the interviewee?

Or consider this wrinkle. It appears that people are willing to pay a great deal more to prevent air pollution in an area with clean air than they will pay to clean up the air in a place where it is polluted, even though the end result would be the same. This principle is called loss aversion (Daviss, 1998). How much clean air is worth depends on what questions are asked and the emotional reference point from which people start.

Despite their shortcomings, valuation methods are critically important to the proper accounting of natural capital. Some valuations can be made more easily than others, but because it is clear that the value of natural systems is not zero, virtually any reasonable attempt at valuation is superior to the failure to valuate.

An Alternative*

Since we know it is necessary to account for natural capital more systematically, and since we have useful (if flawed) tools, such as contingent valuation, to assign values not specified by markets, are we not well on the way toward solving the problem of figuring out how to live sustainably?

Perhaps to an extent. But to proceed much further down this path is not for the faint of heart, for it is extraordinarily complicated to carry out the kinds of detailed evaluations required to even begin capturing all the values of natural capital. This is the inevitable result of insisting on the importance of ecological systems linkages between human economic activity and the functioning of the natural world.

For example, consider three ways to assess the value of the oysters living in the Chesapeake Bay. Within the conventional economics framework, the focus is on the direct value of the oysters. Their value is assessed as what the commercial fishermen, processors, and consumers received or paid for the oysters, multiplied as the exchanges circulate through the economy. From this viewpoint the dollar value, right off the boats, of all the oysters harvested in the bay in 1992 was $2.5 million. Business sales involving the oysters were $7.5 million and related household income was a little over $13 million. About 500 jobs were involved.

Environmental economics, a subfield of conventional economics, still looks at direct values but would expand the previous valuations to include non-use values, such as the value to people who fish for oysters as recreation, the value to people don't fish for or even eat oysters but derive some satisfaction from knowing they are there in the bay, and the value to people who want to leave oysters for their descendants. This complicates the research task by requiring additional surveys of households to find out what they would be willing to pay for those benefits.

Ecological economics casts a still wider net, by examining the role of oysters as part of a vast, dynamic ecosystem that humans also benefit from, completely apart from their feelings toward shellfish or even the Chesapeake Bay:

---

* This section is based extensively on King, 1994, in Jansson et al., 1994, *Investing in Natural Capital: The Ecological Economics Approach to Sustainability.*

The oyster population of the bay, before it was targeted by commercial fishermen, filtered all the water in the bay every three or four days, removing excess nutrients and maintaining extremely high water quality and unmatched biological productivity and diversity. The current oyster population of the bay is 1 percent of this historical level, takes 300 to 400 days to filter the same amount of water, and cannot keep up with the increased nutrient load from expanding agricultural and residential development in the watershed. An ecological economic analysis of oyster values might focus on the resulting buildup of nutrients in the bay that lead to frequent algae blooms and localized fish kills, or the multi-billion-dollar federal and state programs attempting to clean nutrients from the bay, or the cost of restrictions on local agriculture and commercial and residential development to control nutrient flows to the bay. This broader focus on oyster value might also include an evaluation of how excess nutrients in the bay have reduced light penetration to submerged aquatic vegetation, resulting in a decline in critical habitat for finfish. The resulting decline and recent closure of important rockfish and bluefish fisheries in the bay, and the loss of associated jobs, incomes, sales and tax revenues in those fisheries might also be included. ...Ecological economic analysis might show that the real economic value of oysters — their highest and best use — is in their natural role as "the kidneys" of the Chesapeake Bay and not as a temporary source of direct income or recreational enjoyment for fishermen (King, 1994, pp. 329, 331).

Such a comprehensive cost-benefit analysis would obviously be lengthy, detailed, and expensive, and that is one of the major problems with it. Another is the fact that the methods used to assess nonmarket values have many weaknesses, including but not limited to the ones mentioned previously. A third problem is that there may be an element of irreducible uncertainty built into the analysis process, which would involve measuring and integrating a host of chemical, physical, biological, and economic variables and estimating multiple probabilities concerning their interactions. The effect of a seemingly simple event such as the release of a hazardous substance would soon become a tangle of branching possibilities and unknowns, with possible outcomes proliferating like streams in a river delta and drawing analysts into a swamp of doubt about the results: "The propagation of uncertainty as information is passed from one module to another in such an analysis could leave the analyst and decisionmakers without any real confidence in resulting measures of costs, risks or benefits" (King, 1994, p. 335). The end result could easily be ruinously expensive paralysis of analysis.

In such cases, it may be possible to rely on another approach. Natural capital can be classified according to its importance to the ecospheric function and the human economy and according to the reversibility of decisions to use or degrade it for economic gain. These two criteria can be used to determine how much institutional control should be exerted to override normal economic forces and thus protect the resource.

Figure 4.5 shows natural capital divided into three categories using the importance and reversibility guidelines. At the upper left (Group I) are resources that are extremely important and which, if damaged and their ecological functions lost, cannot be restored by human intervention or natural processes. This group would require the firmest constraints on use, such as outright prohibitions or the establishment of protected sanctuaries. At the lower right are resources of relatively lesser importance and high reversibility (Group II). Use of this group could be left to the choices of markets and individuals, provided that all costs, external as

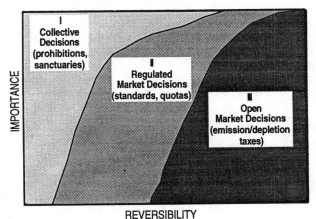

Figure 4.5　Public intervention to protect natural capital. Source: King, *Investing in Natural Capital*, p. 336.

well as internal, are carefully accounted for. The third group consists of those elements of natural capital that lie in between in terms of importance and reversibility. They would not require the very strict controls of Group I but would require more protection than Group II, i.e., some form of conscious management using a mix of standards, quotas, open/closed seasons, and so on. Although further careful research would be needed to classify elements of natural capital into one of the three groups, this scheme acknowledges that some decisions to protect or conserve natural capital can be made primarily on the basis of ecological understanding of its importance, apart from the results of any formal benefit-cost analysis. It also acknowledges that social and political factors should be considered as well.

Both of these approaches — benefit-cost analysis and importance-reversibility classification — need further development and refinement. However imperfect, they represent necessary steps toward properly accounting for natural capital in all its forms as well as creating both a clearer vision of what it means to us and getting a firmer grip on its management. That process, in turn, forms the foundation of the effort to invest in natural capital and thus restore the means whereby we can ensure our own well-being and that of future generations. Section III discusses the problems of investing in natural capital and suggests policies that could encourage it.

MANAGING NATURAL CAPITAL FOR SUSTAINABILITY

III

Chapter 5

Investing in Natural Capital: Incentives and Obstacles

Infinity is ended, and mankind is in a box;
The era of expanding man is running out of rocks;
A self-sustaining Spaceship Earth is shortly in the offing
And man must be its crew — or else the box will be his
coffin!

— Kenneth Boulding
The Ballad of Ecological Awareness

Sections I and II tried to establish the importance of natural capital, the intellectual roots of our neglect of it and why that neglect must not be allowed to continue. The pages that follow discuss some of the reasons for investing in natural capital and then review some of the obstacles to such investment and policies that might encourage it.

The Nutshell Case for Natural Capital Investment*

Conventional economics sees the world as a flow of exchange value between households and firms. The ecosphere (natural capital) is hardly to be found. Ecological economics, in contrast, sees the world as a materially closed, finite, nongrowing ecosystem, within which the human economy is an open subsystem that depends for its very existence on the viability of the ecosphere. The economy — human beings, their machines, buildings and other artifacts, and their economic activity — must draw useful materials from the ecosphere and return wastes to it. The economy is now vast, by one rough measure occupying 40% of the ecosphere. In one doubling time — a generation or so — the ecosphere could be roughly 80% full. That will leave little room for the natural capital necessary to provide us with indispensable life-support functions, much less the resources necessary for economic production. The limiting factor in production is now natural capital, not manufactured capital.

The course of development traditionally prescribed by economics can be summed up as perpetual growth to solve all economic problems. This is not sustainable due to the excessive drawing down of natural capital stocks in order to allow current consumption. Many observers fear that this depletion of natural capital stocks is approaching, or has already reached, critical levels. Therefore, it is essential to stop depleting natural capital and then to begin to restore it, so as to reduce the threat of ecological disaster at least until we have a much better understanding of the importance to a viable ecosphere of the various elements of natural capital.

Keeping natural capital intact, and ultimately investing in it, means maintaining its stocks separately from those of manufactured capital. Conventional economics tends to assume that the two forms of capital are largely substitutable. This implies that it may be economically "optimal" under some conditions to harvest or deplete natural capital completely, so long as the proceeds can be invested in manufactured capital that yields a higher return on investment (income). Ecological economics, on the other hand, argues that natural capital and manufactured capital are

---

* Except as noted, this section is based on Daly, 1994, in Jansson et al., 1994, *Investing in Natural Capital: The Ecological Economics Approach to Sustainability.*

essentially complements, not substitutes. Therefore it is important to maintain natural capital stocks intact independent of manufactured capital stocks. This is called strong sustainability.

If natural capital is now the limiting factor in human economic activity, then clearly we must devote more time and attention to making the most of it and increasing its supply for the future. Continuing to emphasize the investment in more manufactured capital simply widens the gap between the two stocks of capital, just as building more fishing boats — when what is needed is more fish — simply accelerates the rates at which overharvested fish stocks are driven further toward total depletion.

What does natural capital investment mean?   Essentially it involves refraining from current consumption in order to restore capital stocks and thus ensure future consumption. To keep stocks at their current levels, it is only necessary to reduce current consumption — the rate of depletion of natural capital — to the surplus produced each year. Natural capital, especially renewable resources, tries to grow. If we consume each year's net growth, the total stock remains the same size. If we consume less, however, the stock grows, and as it grows it becomes capable of producing larger income flows in the future. If the world's population were stable, we could leave natural capital stocks at their current size and be assured of a stable per-capita level of consumption. But since the population is growing, ensuring even a hope of stable per-capita consumption of the products of natural capital in the future means allowing stocks to grow as well.

The case of nonrenewable natural capital, especially fossil fuels, is somewhat different. Nature creates these resources so slowly that there is really nothing we can do to regenerate the available stocks in any meaningful period of time. Our only option is to use them or leave them in the ground. In the case of fossil fuels, there may be compelling climate-related arguments not to use them at all, or at least to use them as little as possible. But if we use them (liquidate the world's inventories of them), we must not make the mistake of calling it income. It is, instead, selling off the inherited capital of the geologic past and consuming the profits in the present. This bubble must burst eventually. In the long run we would be far better off investing some portion of the proceeds in renewable alternatives, so that the transition leads to a sustainable future rather than a return to pre-industrial standards of living.

How does one leave natural capital alone, so it can regenerate?   By reducing throughput. There are two ways to do that. Recall that Impact = Population × Affluence × Technology. One focus, then, should be on investing in population growth reduction. This applies globally, not just to the developing world, since the high consumption rates of the developed world mean that every baby born there is likely to have a greatly disproportionate effect on natural capital stocks. The means and techniques necessary to achieve, first, lower growth, then zero growth, and eventually population reductions, will differ from one place to another, but the need is pressing everywhere.

The second investment approach to reducing throughput is to increase the efficiency with which each unit of throughput works to satisfy human wants and needs, so that fewer units are needed.

To understand this, we need to remember that humans have always converted natural capital to manufactured capital in order to make better lives for themselves. The harnessing of fossil fuels merely enabled us to do it much faster and on a much bigger scale. Harvesting a few reeds to make a basket is not different in principle from mining iron ore and turning it into 80,000 tons of steel for an aircraft carrier or a fleet of automobiles.

The issue is not whether to convert natural capital, but rather how much benefit we gain from the manufactured capital *vs. how much we lose by sacrificing the natural capital and its services to make the manufactured capital.* Long ago, when there were few people and the economy was relatively small, natural capital could be freely converted (and its services sacrificed) without posing much danger to the ecosphere. That is no longer true, hence the need for greater efficiency to reduce throughput.

There are four aspects of this efficiency:

- *The service efficiency of manufactured capital.* How well and efficiently do machines and products do what they're supposed to do? How efficiently are resources divided among various economic uses? And how equitably are the fruits of economic activity distributed among the population? — since, up to some point, total usefulness or want-satisfaction will be increased by more even-handed distribution (see the brief discussion of income distribution in Chapter 4, page 86).

- *The durability of manufactured capital.* The longer a machine or building lasts, the less often it must be replaced, thus lowering throughput. This implies that products should be designed to be durable, easy to repair, and easy to take apart and recycle when their useful lives are over.

- *The growth efficiency of natural capital.* If renewable resource stocks expanded faster, more would be available for immediate consumption without affecting total stock levels. Genetic engineering may or may not be able to expand the biological limits on renewable resource growth without causing other problems, but there are ways we can optimize our use of the differences that already exist among various species.

- ***The service efficiency of ecosystems.*** Natural capital provides many different products and services. Exploiting it for one product or service often means giving up some of the others. Forests cut down for lumber can't absorb carbon dioxide, prevent erosion, or help preserve biotic diversity by providing habitat. Our objective should be to minimize these ancillary losses even as we use natural capital as resources for production.

Finally, it is important to note that even if we struggle relentlessly to maximize all four of these efficiency dimensions, continued throughput growth will still bring us to the point at which the losses of goods and services from sacrificed natural capital outweigh the gains from converting it to manufactured capital. At that point, it is useless to continue, since net total welfare would have begun to decline. The only option would be to return to the I=PAT equation and lower throughput by reducing population or standards of living.

It is difficult to exaggerate the urgency of the need for investing in natural capital. It was not long ago that the domestic crisis *du jour* was our declining infrastructure, i.e., our crumbling highways, leaky water systems, and overcrowded airports. Our competitiveness and our standard of living were at risk, it was said. That crisis may have faded from the headlines, but the crisis of natural capital infrastructure is worsening. For natural capital is the biophysical infrastructure of the entire human niche (Costanza et al., 1997) and the foundation of all human economic activity. Investing in natural capital is investing in infrastructure at the grandest and most critical scale.

Why We Ignore Natural Capital

Since investing means, in part, doing without current consumption, we need not look very far in this consumption-oriented world to find reasons for our cavalier attitude toward natural capital and the need to invest in it. We have already mentioned some of the key reasons, such as the fact that during most of the human tenure on the planet resources were not scarce, except perhaps in particular places at particular times. Economics as a discipline was invented near the tail end of that era. Since mainstream economics sees natural capital as endlessly abundant and/or highly substitutable with other forms of capital, it has institutionalized the belief that production requires only inputs of labor and capital to proceed.

In practical terms, the effect is that individual resource users are often not made to face the full social costs of their private decisions about natural capital use (Barbier, 1994). (See discussion of externalities in Chapter 6.) The reasons include the failure of markets to adequately reflect environmental values (such as the nonuse values, conservation values, existence values, and option values discussed in Chapter 4) and the failure of institutions to guide markets by setting up proper structures and ensuring adequate information flows. In many instances, individuals do not even have to pay the full *private* costs of their use of natural capital, as in the case in the United States of federally subsidized below-cost timber sales, cheap mining claims, and fees for grazing cattle on federal lands that are set too low to encourage sustainable use.

This brings us to social traps (Costanza, 1987). A good case can be made that a rational system for managing the Earth should acknowledge that the goals of long-term global ecological and economic health and sustainability will often be more critical than local, short-term goals for economic growth or the goals of private interests. But our institutions and incentive structures are set up to address primarily these short-term, private interests. And since people, companies, and countries consistently pursue private self-interests and frequently are not checked by the institutions, they thus tend to undermine the larger goals. That is, the institutions and incentive structures set social traps.

A social trap (see Figures 5.1a–5.1d) exists when the prospect of a local, immediate payoff encourages behavior inconsistent with long-run and/or broader goals. Pesticide overuse is an example. No farmer has a real long-term interest in encouraging the evolution of resistant strains of bugs that can only be killed by heavier applications of ever-more lethal chemicals. However, all farmers have an immediate interest in protecting their crops and raising yields in any given year. So, they keep spraying.

Intervention may be required to eliminate social traps. There are several approaches, among them:

> ***Education***. People can be systematically informed of long-term consequences of their short-term-focused behaviors. We tell students, for example, that smoking causes cancer, drinking

Figure 5.1a Social traps as misleading road signs, and some potential solutions (Figures 5.1b–d). Source: Costanza (1987).

causes birth defects, and unprotected sex can transmit AIDS. To varying extents, such warnings can help reduce the socially borne costs of cancer, institutional care, and prolonged treatment of terminal illness. On the other hand, education about traps is costly, time-consuming, unlikely to reach everyone about all the traps they may be in, and often ignored or forgotten.

Figure 5.1b   Educating against social traps.

Authority. Governments can ban social-trap behaviors. In the United States, one of the most famous examples was Prohibition, which has been said to have: (a) worked; (b) failed; or (c) worked but cost too much. The high cost is not disputed and

Figure 5.1c Banning social trap behavior.

that is typical of government bans of individual behavior with very strong short-term incentives, since the necessary system of police, jurisprudential machinery, and prisons is inevitably a great drain. There are also important nonmonetary costs, such

Figure 5.1d   Turning social traps into trade-offs.

as the loss of freedom. Short of bans, governments can regulate, which is essentially banning behavior outside a certain range. The same criticisms apply, but to a lesser extent. Other sources of authority, such as churches or a general sense of a social contract among citizens that helps define expected and accept-able be-havior, can also be effective, at least in stable societies where people are much alike. They may work much less effectively in rapidly changing societies with a broad mix of ethnic and cultural flavors.

Conversion to trade-offs. Turning a trap into a trade-off means bringing some or all of the long-term costs into the present, so that they are immediately apparent to people and can help guide their decisions.

An example of how this works comes from a game used to model how common property resources can be managed (Edney and Harper, 1978, cited in Costanza, 1987). A stack of poker chips is used to represent some common renewable resource, and players try to maximize their use of the resource over time. Players can take up to three chips apiece during each round, but since the "resource" is renewable, the stack of chips is replenished after each round according to how many chips are left in the stack. That is, just like a stock of fish or trees, the larger the capital stock, the more surplus it can produce. Players would seem to have a strong short-term incentive to grab the maximum three chips per round, but if everybody does this at every turn, then the pool shrinks quickly and soon produces little surplus. Taking fewer chips means that the surplus is larger and can keep the total stock stable or even allow it to expand, thus allowing all players to acquire more chips over the long haul. One way to encourage the taking of fewer chips is to "tax" each acquisition by taking back all chips above, say, one. This eliminates any incentive to take more than one per round (unit of time), which might be what the system can supply over the long term. This approach is analogous to the natural capital depletion tax discussed in more detail in Chapter 6.

Command-and-Control Regulation*

The early and still-dominant response to pollution and environmental degradation in the United States and many other places is command-and-control regulation (CAC). CAC, as the name suggests, relies on the exercise

---

* This section draws extensively from Section IV of Costanza et al., 1997, *An Introduction to Ecological Economics,* and other sources as noted.

of governmental authority (especially federal authority) to avoid the social traps of natural capital destruction. To date, the most prominent U.S. efforts in this direction have sought to prevent the depletion of the waste absorption capacities of natural capital by controlling pollution of the air and water. Other efforts, such as the government's work to mediate a settlement between timber interests and environmentalists over old-growth forest in the Pacific Northwest and thus help preserve biodiversity, are more recent.

CAC approaches would seem to have the benefit of straightforwardness: identify a polluting practice, pass a law against it, and write a regulation saying exactly what may and may not be done. And CAC can claim some notable successes. In the United States, for example, the decision to ban lead in gasoline and paint has markedly reduced ambient levels of this toxic metal. Such clear-cut gains are less numerous than we might wish, however, and widespread application of CAC to attack all manner of environmental problems has demonstrated how complex and politically contentious it can be. The long reliance on CAC seems to stem partly from its conceptual simplicity and familiarity and partly from the fact that we have always done things this way. Now, of course, the existence of large regulatory bureaucracies and legions of corporate lawyers specializing in regulatory minutiae creates substantial vested interests in continuing to use this approach.

Nevertheless, a case can be made against CAC. Empirically, it often fails. The 1970 Clean Air Act and later amendments, for example, instructed the Environmental Protection Agency (EPA) to write clean-air standards (called National Ambient Air Quality Standards, or NAAQS) for six so-called "criteria pollutants:" nitrogen oxides, sulphur oxides, carbon monoxide, particulate matter, lead, and ground-level ozone. Emissions of these pollutants are mainly the result of industrial activity and burning fossil fuels in vehicles. EPA prescribed specific technological approaches (such as catalytic converters and fuel efficiency standards) for cars and trucks, and required every state to come up with detailed plans to ensure industry compliance with the NAAQS. Deadlines were set for compliance and penalties spelled out for noncompliance. Under the present arrangement, auto manufacturers can be fined for violating emissions or fuel-efficiency standards. In general, the states are responsible for ensuring that all areas within their jurisdictions are in compliance with the NAAQS. Pressure on the states is supposedly created by the federal threat to withhold money for highway construction, maintenance, and other programs in the event of failure to meet the standards.

Has this elaborate plan worked? As already noted, lead levels dropped sharply, but compliance with the rest of the standards has proved elusive.

The deadlines have been postponed several times when major cities and regions with strong political representation in Congress were unable to deploy or enforce sufficiently rigorous measures to achieve compliance. EPA publishes an annual list of cities that exceed the NAAQS, usually for carbon monoxide and ozone, and the list remains lengthy.

From the economist's point of view, CAC is frequently inefficient in the sense that it does not necessarily lead to "optimal" use of resources, i.e., that pattern of use which yields the best balance of benefits and costs. The economist argues, with good reason, that markets are generally more efficient at meeting people's needs than are bureaucracies doggedly issuing mountains of regulations (recent U.S. environmental laws have tended to run into the hundreds of pages, and each page of law typically requires 10 to 20 pages of implementing regulations). From ecologists' viewpoint, CAC rarely goes far enough; they often see economic efficiency as secondary to thoroughly protecting ecosystems and their living constituents. CAC also suffers from these additional disadvantages:

- The amount of technical and company-protected information that regulators must have to be effective is great and often simply unavailable.

- Even when the necessary information is available, the scope and scale of the oversight required means that monitoring and enforcement can be extremely expensive.* Consequently, the unit costs of pollution reduction are high. In the absence of offsetting charges or taxes against industry, the costs are borne by taxpayers, not by the polluting industries and their customers.

---

* The overall costs of CAC regulation are the subject of debate. Studies using computer simulations of the economy suggest that, although environmental regulation may retard GNP growth, the effect is relatively small (Prince and Gordon, 1994). For example, one study covering the period from 1973 through 1982 estimated that antipollution regulation reduced measured national output growth by 0.09% per year. A second study for the period from 1973 through 1985 estimated GNP growth to be 0.19% lower than it would have been without the regulations. On the other hand, a third study that used statistical methods to analyze historical data from 1973 through 1989 found no evidence of GNP shrinkage and even concluded that states with strong environmental protection programs had the highest levels of job growth and creation during the period.

- The sheer volume and complexity of the tens of thousands of pages of regulations means that regulations are often ignored — deliberately or accidentally — without penalty.

- CAC regulations tend to establish specific, statutory emissions limits or require the use of particular technologies that are expected to reduce emissions by predictable amounts. Once a company or industry has complied with those requirements, it has no incentive to reduce emissions any further.

Finally, the CAC system, being based on legal tradition, presumes innocence (compliance with emissions limits) unless guilt can be proved. However, the fate and effects of pollutants are highly uncertain in many cases, which means that proving a violation of regulations can be quite difficult. The Superfund program illustrates the point, although its purpose is to clean up old toxic waste dumps and spills rather than prevent new pollution. Billions of dollars have been spent trying to identify so-called Potentially Responsible Parties (companies or individuals who may have contributed to the accumulation of wastes at various sites around the country) in order to assess them for their share of cleanup costs. Since those costs are in many cases quite substantial, PRPs have strong incentives to fight the assessments. Most of the Superfund budget to date has gone to pay lawyers and consultants rather than to actually clean up the contaminated sites.

Incentive-Based Systems*

There remain many circumstances when CAC is a useful approach, but its shortcomings suggest there is plenty of room for alternatives that are less costly and more efficient. One class of such alternatives are incentive-based (IB) systems, which seek to build more economic efficiency into the process of controlling pollution and environmental degradation.

IB systems rest on the following general principles. Firms attempt to maximize their profits and minimize their costs of operation, often through the time-honored way of using "free" open-access resources, e.g., water and air, as places to dump wastes. This represents unpriced or unpaid-

---

* This section draws extensively from Section IV of Costanza et al., 1997, *An Introduction to Ecological Economics*, and other sources as noted.

for damage to, or depletion of, natural capital. In order to prevent or correct these damages, firms must be required to pay a price.

Here, some would argue that the proper price is whatever price is high enough to force firms to emit no pollution whatsoever. This is a sticking point, for all human economic activity creates wastes, or pollution. (Recall from Chapter 2's discussion of entropy that evolution is pollution.) So while nobody likes pollution *per se*, we see it as an acceptable tradeoff for the benefits, in terms of goods, services, and quality of life, that we get from economic activity. The real issue is how to determine the optimal tradeoff between the ecological costs of economic activity and its quality-of-life benefits. (We argue in this book that the ecological costs of economic activity are far graver than is generally supposed and thus need to be rapidly and carefully addressed, not that humans can or should live without incurring any ecological costs at all.)

For our hypothetical firm, low levels of output produce low levels of pollution and thus low levels of damage to the environment. If output is modest enough, pollution levels will be so low that the environment can assimilate the wastes without any problem. However, as output and pollution levels rise, so does environmental damage. As the waste-assimilation capacities of the environment are approached and then overwhelmed, environmental damage increases sharply.

Another process is at work at the same time, the one that describes what it costs to achieve the ecological benefits of controlling pollution. If our firm's output is high and it is subject to no environmental controls, then the cost of achieving significant reductions in pollution (and the associated ecological benefits) is low per unit of production, because there are generally some easy and simple options for stopping the worst of the pollution. These do not cost much, so the ecological gains are high for every dollar invested in controls, at least at first. The more pollution reduction is sought, however, the greater becomes the investment required to achieve another unit of reduction. At the very end of the curve, it costs an astronomical amount to achieve very tiny reductions (and, of course, it is impossible to eliminate pollution completely, at any cost, and still have any output). But long before then, society probably would have decided that such gains were not worth the cost, since the huge sums of money spent on pollution reduction would be unavailable for other valued things.

This is how we can wind up talking about the "optimal level of pollution," even though it is a concept that some find repellent. Many people argue that it is the immorality of pollution that demands CAC regulation of it, in much the same way that society flatly forbids murder on moral grounds rather than trying merely to set up an incentive structure to keep murder at "optimal" levels. It is sometimes asserted, along these lines, that the incentive-based system of emission charges is a license to pollute. Ignored by this argument is the fact that a CAC structure also creates a license to pollute, and it's free. Both systems can control pollution, but IB structures levy charges on polluters (therefore, by extension, on their customers) and thus have the advantages of: (1) creating incentives to develop new technologies and otherwise reduce pollution; and (2) generating funds. CAC structures create only limited pollution-reduction incentives and actually cost taxpayers, not polluters and their customers. In that sense, to switch from a CAC system, with its no-charge emission privileges, to an IB system is to take wealth from polluters and give it to the public.

IB systems are well established to be much more efficient economically than CAC systems; studies of air-pollution control options in various U.S. cities suggest that CAC approaches would cost up to 14 times as much as IB systems for the same level of air quality. One of the chief reasons is that IB systems take advantage of the fact that different firms will have different pollution-control costs. Under a regulatory regime, with some absolute cap set on, say, per-factory emissions, an efficient firm with up-to-date equipment might meet the standard easily, with plenty of room left over for further improvements — but no incentive to pursue them. Under an IB system, that firm can save money by further reducing emissions, and society in general comes out ahead because it gets more pollution control at less total cost. Alternatively, when an IB system incorporates marketable pollution permits, firms have an incentive to develop better pollution-control technologies because they can use them to reduce emissions and then sell their unused permits to other firms using more polluting technologies.

Of course, there are some flies in the IB ointment. If pure incentive-based systems were completely flawless, there would be a lot more of them. The problems are of several types. For one thing, as noted above, there are strong vested interests in the traditional, regulatory way of doing things. Polluters generally resist the idea of paying for a necessary service (waste assimilation) that has traditionally been free.

But not all the blame can be laid at their door. Bringing the market into the picture forces us to confront the shortcomings of markets, one being that markets will not necessarily address the overall scale of pollution. That is, while it might be possible to achieve optimal economic efficiency by structuring things so that polluters are encouraged to reduce emissions to the lowest economically achieveable levels, the collective total of emissions might still be too large for ecosystem integrity.

Another problem is that markets suffer from failures, such as lack of perfect information among all participants, which impair their theoretically achieveable efficiency. A market-based system might also function efficiently when viewed at a national level yet still impose unacceptable levels of pollution or environmental degradation upon specific individuals or communities. In some cases, such as emissions of an extremely toxic substance, the only acceptable level of emission might be zero, in which case an outright ban, not the optimal market-set level, would be the policy of first choice.

Finally, like CAC systems, market-based systems do not avoid the very difficult scientific problems of figuring out what levels of emissions result in what levels of ambient concentrations, what effects those concentrations have, and how to value the resultant damages in order to arrive at a figure for pollution charges. The multitude of pollutants, emission sites, affected species and ecosystems, and possible interactions among different pollutants threaten to overwhelm the approach in a chaos of detail. Best approximations and careful, educated guesses are the order of the day.

Chapter 6

Some Investment Strategies*

Markets are only meant to allocate resources in the short term, not to tell you how much is enough, or how to achieve integrity or justice. Markets are meant to be efficient, not sufficient; greedy, not fair.

— Amory B. Lovins

Since both command-and-control and incentive-based approaches to investing in natural capital have shortcomings, there is much to be said for combining them in complementary ways. The U.S. Clean Air Act Amendments of 1990, for example, incorporated a CAC measure by setting a nationwide cap on total allowable emissions of sulfur dioxide, which

---

* This section is based on Costanza et al., 1997, *An Introduction to Ecological Economics*, and other sources as noted.

is produced mainly by coal-burning electric power plants. However, the law also introduced a system of marketable or tradeable emission permits, so that the market could be used to allow the maximum of pollution reduction at the least cost.

This sort of legislative instrument can be extremely useful. This chapter describes some other general proposals that also attempt to combine in various ways the better features of CAC and IB systems to create structures that protect and encourage investment in natural capital. It also briefly discusses property rights regimes and, in that context, common property systems, which are alternatives to conventional Western ways of viewing resource ownership and access that have proven successful in many places and times in managing resources sustainably. The chapter concludes with snapshots of the efforts of two nations (The Netherlands and Costa Rica) to bring their differing economies into line with the demands of sustainability.

It is important here to note in passing that there is a compelling need for influential institutions, as well as strategies, that work to preserve natural capital by addressing the need to constrain the overall scale of the economy in order to achieve sustainability. Not surprisingly, few such institutions exist. Given the degree to which the ideology of growth and the sense of its utter normalcy are embedded in our culture, what is surprising is not that there are so few institutions for controlling economic scale, but that there are any at all. Institutions with an interest in promoting growth and the expansion of scale are legion: their names are Kiwanis, Rotary, and Lions. Other names include transnational corporation, chamber of commerce, department of trade and industry, agency for international development, booster association, export marketing administration, and so on. Such organizations are globally ubiquitous. At the international level, agreements such as bilateral trade treaties, the North American Free Trade Agreement, the General Agreement on Trade and Tariffs (and its institutional offspring, the World Trade Organization), and the Multilateral Agreement on Investment are designed expressly to encourage global expansion and integration of the human economy.

Opposing this array of growth machines we find mostly local zoning boards and land-use planning commissions. In the United States and elsewhere, a scattershot collection of laws and government agencies (e.g., the Environmental Protection Agency, the Endangered Species Act, the system of national parks and forests) addresses issues indirectly related to the scale of economic activity. Internationally, a few key agreements (such as the Global Climate Change Convention) embody the world's grudging belief that there are at least a few dimensions of economic growth that probably ought, in the name of prudence, to be controlled.

There is no national Economic Scale Administration or, at the international level, a United Nations Global Economic Scale Program. Perhaps there never will be. However, the investment strategies discussed below, were they to be adopted, would very likely help to control scale, at least locally or nationally. Aggregated, such control would be a step toward global control of scale as well. The implementation and administration of the strategies might lead to the creation or evolution of the appropriate institutions. Though not necessarily the most efficient means of creating a system to cope with the problems of limiting scale, it would be far superior to ignoring those problems altogether.

Green Taxes

Green fees or taxes are an expression in policy of the polluter-pays principle: whoever causes pollution should pay for it. They represent the application to environmental problems of Pigovian taxes (named for A.C. Pigou, the economist who proposed them in the 1920s). Pigovian taxes are intended to address an important form of market failure called externalities. These are the unintended effects of production or consumption by firms or consumers that are not mediated by markets and thus impose costs (or deliver benefits) that are not reflected in the price of the transactions. These unintended consequences often affect people who are not participants in the transactions.

Externalities can be positive or negative. Suppose a crime-conscious suburbanite installs powerful floodlights under the eaves of his house and leaves them shining on his back yard from dusk to dawn. If the lights illuminate the adjacent yards as well, the neighbors may benefit from this deterrent to burglary even though they haven't helped the owner buy and install the lights or pay for the electricity to run them. This would be a positive externality (and the neighbors would be "free riders"). On the other hand, perhaps the neighbors think the owner watches too much television and is simply paranoid about crime. Maybe they resent the glare because it keeps them awake at night and makes their yards look like prison compounds. They are then the victims of a negative externality: the price of the lights does not include the cost of the neighbors' insomnia and the loss of ambience.

The environmental pollution and degradation incident to economic activity frequently produce negative externalities. That is, they impose environmental costs not reflected in the prices charged for the goods and services produced. Factory and power-plant smokestacks, for instance, emit gases and small particles that help create acid rain, which in turn contributes to forest dieback, the acidification of fresh-water bodies, and the corrosion of building materials. But consumers do not pay for these losses when they buy the clothing, kitchen utensils, or miniblinds produced by those factories (although their tax dollars may be spent on government-sponsored mitigation programs, whether they bought the products in question or not). As we saw in Section II, hydroelectric dams can decimate fish stocks and even lead to extinctions, as they have in the Pacific Northwest, but electricity ratepayers will search their bills in vain for a "biodiversity loss surcharge." Industrial agriculture may deplete aquifers and dissipate topsoil, but those costs aren't fully felt (yet) in the price of corn or wheat. Commuters stuck in a traffic jam on the Santa Monica Freeway, their idling motors pumping out smog-forming chemicals, help increase the incidence of respiratory problems among children, asthmatics, and the elderly — but the costs are paid via the taxes and health insurance premiums of drivers and nondrivers alike, not through gas prices.

Addressing externalities means getting the price right. The market prices consumers pay for everyday goods and services are often wildly misleading. The use of motor vehicles in the United States, for example, costs an estimated $300 billion per year more than vehicle operators pay directly (MacKenzie et al., 1992; see Figure 6.1). Economists have long recognized that achieving maximum economic efficiency requires that the true and full costs of any economic output must be accurately reflected in the prices paid for that output.* Only when natural capital is viewed as a free good (the empty-world mindset) can anyone even think of justifying its use as an input and then charging nothing for its depletion. Failing to account for the costs of resource depletion, biodiversity loss, and pollution means that the market cannot set prices properly. That denies both producers and consumers accurate information about the real costs of their economic decisions.

Green taxes can remedy this problem in many cases, at least in theory. (Getting people to pay the real costs thus imposed, when they have long been accustomed to a partially free ride, is another issue. Any plan to

---

* Getting the price right is also the way that conventional economics views the obligation of the system toward maximizing economic efficiency. The asthmatic with smog-aggravated shortness of breath can be treated and even compensated for the symptoms, but even if the motorists whose driving leads to the smog that aggravates the asthma could somehow be assessed their share of the cost, the question of whether it is right remains. If asthmatics had perfect information and freedom of movement, as markets assume, most would probably not choose to live in smog-laden areas even if all their medical costs were picked up by someone else.

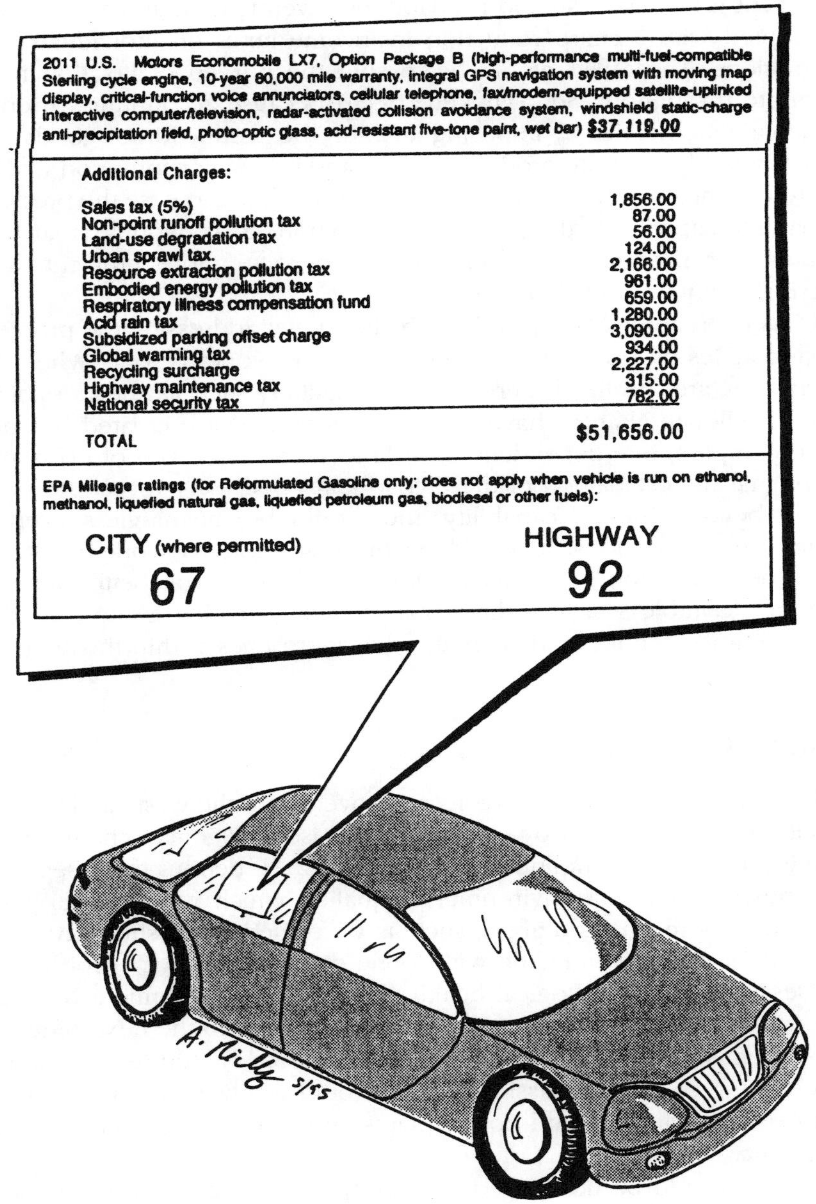

Figure 6.1 What an auto price sticker of the future might look like if all external costs of driving were reflected in the price.

implement green taxes should probably be revenue-neutral; i.e., it should be coupled with a sharp reduction in income taxes on individuals and corporations. The effect would be to shift the burden of taxation from income to throughput. See Hawken, 1993.) Because accounting for environmental externalities by imposing a tax on the production of goods and services would raise both production costs and prices, consumption would tend to decline, thus reducing throughput and helping to ease the pressure on natural capital. Further, higher production costs would create an incentive for producers to find ways to become more efficient, also lowering pollution and depletion.

It has been argued (Daly and Cobb, 1989) that addressing the problem of externalities by such means as green taxes is workable only when the externalities are localized. Pervasive externalities, such as the possibility of global climate change, have countless facets and are created by such a vast, complex, interdependent, and difficult-to-assess web of economic actions that the calculation of the myriad taxes necessary to get the price right is beyond human capability and would be meaningless even if possible. In these cases, it would be more sensible to work backward from a determination of the maximum possible *scale* (or, better yet, the optimum desirable scale) of the global economy, which is more readily definable, and then let markets figure the right prices within those limits.

Graded Ecozoning

Graded ecozoning is a way to address the overuse of the waste assimilative capacity of natural capital by combining the use of Pigovian charges with administrative caps on pollution (see Figure 6.2). It defines three separate categories or ranges of environmental quality, which would be applied to ecologically meaningful areas, such as watersheds or airsheds. An area would be designated as falling within one or another range according to an assessment of its ecological health. As changing economic activity in an area shifted it from one range of ecological health to the next, different levels of emissions would be allowed and enforced by the use of successively tighter controls. Emissions limits would apply to both stationary ources such as factories and power plants and to mobile sources such as automobiles.

An area would be designated a *property-right zone* if emissions there fell within the lowest-emission range. Emissions would have to be so low that ecological damage was too slight to be measured or observed, or at least too low to reduce ecosystem productivity. Emitters would not be taxed or charged for emissions that did not cause significant damage and did not accumulate in significant concentrations. In economic terms, at this emissions level the costs of constant monitoring would exceed the costs of any damage. The costs of monitoring in property-right zones would fall on taxpayers in general.

An area would be designated an *incentive zone* if it were clear, by ecological criteria, that emissions had caused environmental damage and were measurably threatening the productivity of the system. Here a charge on pollution would be imposed that was heavy enough to prevent total emissions from growing to the point of causing nonoptimal damage, i.e., the point at which the benefits of production would be less than the total value of the environmental services lost. Emitters would thus have an economic incentive to reduce emissions to at least optimal levels and even to those levels at which no damage occurs. Motorists, for example, might be encouraged to stay out of an area of high ozone or carbon monoxide concentrations by requiring the payment of tolls on stretches of road within the area.

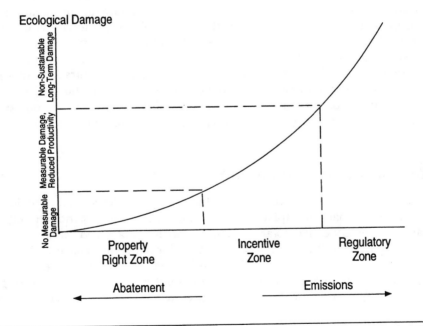

Figure 6.2   An ecological economic approach to pollution control. Source: Cumberland (1994).

The need for a third, *regulatory zone,* is based on the fact that some polluters would be willing and able to pay even heavy emission charges and continue emitting pollutants at high and damaging levels. When it became clear that such emissions were approaching levels that threatened irreversible and unsustainable damage, a regulatory zone would be declared and bans on further increases imposed.

Like other systems for controlling pollution optimally, graded ecozoning would require an effective nationwide system for monitoring ambient concentrations of pollutants as well as monitoring systems and criteria for gauging the ecological health of an area. All of these are either in use or achievable now.* A system much like this is in operation in The Netherlands, where progressively higher charges on "emissions" of livestock manure are levied as manure output per hectare increases. At the high end (what would be called the regulatory zone), although there is no regulatory cap on output, charges rise to twice the incentive-zone rate.

Among the advantages of graded ecozoning are that it doesn't penalize economic activity that does not cause ecological damage. It further requires polluters to pay for the damage they cause and creates an incentive structure for polluters to reduce emissions to more tolerable levels. The system also includes features that enhance its political acceptability, i.e., it secures property rights in areas of no ecological damage and guarantees their marketability. Another plus is that, by employing criteria that emphasize ecological health and ambient concentrations rather than emissions per se, it focuses on the variable that directly affects health. The complex interrelationships between emissions and ambient concentrations of various pollutants would not need to be known in detail. Finally, graded ecozoning avoids the charge of being an unlimited license to pollute by retaining the regulatory ban on emissions beyond a certain level.

Natural Capital Depletion Taxes**

Under the principle of strong sustainability it is considered vital to maintain the stocks of natural capital intact in and of themselves, not merely to maintain the collective total of natural and manufactured capital stocks. This may be called the prudent minimum for achieving sustainability. To achieve this, it would be useful to have an economically efficient instrument to encourage the conservation of natural capital, i.e., lower throughput. The natural capital depletion tax, which would heavily tax all

---

* For more information on ecological health assessment, see *Ecosystem Health: New Goals for Environmental Management*, by R. Costanza, B. Norton, and B. Haskell, Eds., Island Press, Washington, D.C., 1992.
** This section and the two following sections are based on Costanza et al., 1997, *An Introduction to Ecological Economics*, and other sources as noted.

consumption of natural capital, is one such possibility. It would work best if combined with a restructuring of the income tax system in order to sharply reduce income taxes, especially on the poor.

Taxing natural capital consumption is the opposite of historical practice, at least in the United States, and would help correct a serious and unsustainable market distortion:

> [The market] is not neutral at all: today's market was constructed over centuries to encourage the consumption of apparently inexhaustible resources. For instance, extractive industries get depletion allowances, tax breaks that lower the cost of using up their deposits of natural resources. If society's best interests now lie in conserving resources, then the existing market works *against* those interests. In that case, those structures that encourage the consumption of resources need to be altered or removed (Gever et al., 1991).

A depletion tax should appeal to a wide variety of groups that see eye to eye on little else, provided other taxes were reduced as mentioned. Technological optimists — those who believe that technology will never disappoint perpetual faith in its miracles — should at least concede that a depletion tax will create technology-driving economic incentives to vastly improve the efficiency with which natural capital is extracted and used. Technological skeptics should like it because it would retard the rate of natural capital depletion and allow it to regenerate (renewables, that is; but heavily taxing fossil fuels would make several alternatives competitive or economically superior right away).

In picking the precise rate at which to tax natural capital, we could wish for better methods of valuation. But it is not necessary or advisable to wait for them. Rough estimates would do for now, and they are achievable. They can be adjusted later as more and better data are gathered and improved valuation methods are developed. It is likely that we are already at or beyond the optimal, sustainable scale of natural capital depletion (see Section II), and measures of some sort are necessary:

> It would be helpful to have better quantitative measures of these perceived costs, just as it would be helpful to carry along an altimeter when we jump out of an airplane. But we would all prefer a parachute to an altimeter if we could only take one thing. The consequences of an unarrested free fall are clear enough without a precise measure of our speed and acceleration (Costanza, 1994, p. 396).

The Link Between Human and Natural Capital*

[T]he economic logic behind dumping a load of toxic waste in the lowest-wage country is impeccable. — Lawrence Summers, Former World Bank Vice President and Chief Economist, cited in *The Economist*, February 8, 1992

One of the most difficult and contentious issues related to sustainability is the global distribution of wealth. Yet, to the extent that poverty, misery, and deprivation in the developing world often seem locked in a destructive embrace with environmental degradation, the sustainable use of natural capital will remain out of reach until ways are found to relieve at least the worst of that poverty. Policies to achieve that end, mainly by spending on improved nutrition, health care, and education, can be thought of as policies to invest in human capital.

The distribution of wealth and income do not concern neoclassical economics. Its emphasis on efficiency and its striving to be a value-free science mean that distribution is seen as a social problem, not an economic one. When pressed, economists have often argued that problems of distribution can be finessed by promoting growth: there's no need to slice the pie differently because we can simply bake

_____
* This section is based largely on Segura and Boyce, 1994.

another one. But inequalities of wealth and power help explain why environmental degradation occurs in the course of economic activity, and why those who benefit are not the same people as those who suffer the costs:

First, since environmental degradation is...an increasing function of consumption and production, the fact that the rich consume more implies that they generate more environmental degradation. Second, since money can be spent to reduce or avoid bearing the costs of pollution, for example, by residing and vacationing in relatively uncontaminated ecosystems, the rich can more readily escape the costs of environmental degradation. Finally, wealth is positively correlated with power, and power increases one's ability to impose negative externalities on others and to resist having them imposed on oneself (Segura and Boyce, 1994, p. 483).

There are at least three ways in which investment in human capital can reduce environmental degradation. First, poor people are often forced by desperation to focus on the near term. If that means farming marginal land or destroying a patch of rainforest for two or three years' worth of crops, there may be few alternatives — even when the poor themselves understand the damage thus caused. Human capital investments can ease some of the heavy pressure to focus only on the immediate problem of survival and permit a longer and wider view with more possibilities.

Second, the poor often suffer from the environmental destruction wrought by the rich simply because they lack the power to resist it:

> In Central America, where very rapid deforestation took place in the 1960s and 1970s, the main cause was the clearing of lands for cattle ranching.... The main beneficiaries...were the large hacenderos and owners of meat-packing plants.... The main losers were poor peasants, who were denied access to previously cultivated as well as forested lands, and for whom extensive cattle grazing provided meager employment opportunities. ...In Southeast Asia, the main cause of rapid deforestation has been logging for tropical hardwood exports. The...beneficiaries are logging concessionaires — who are often military officers and the political cronies of top government officials.... Once again, the...losers are the poor, including displaced forest dwellers (often ethnic minorities) and downstream peasants whose crops depend on the forest's hydrological "sponge effect" (Segura and Boyce, 1994, pp. 484-5).

Human capital investments in the poor can improve their capacity to understand the process of environmental degradation and to mount political and economic resistance to the destructive and exploitative actions of the rich. Environmental education in particular can build on the poor's indigenous knowledge of the local ecology, improving their knowledge of environmental costs.

Finally, improvements in health care and education often lead to declining birth rates, as women find opportunities for employment outside the home, as access to birth control improves, as the need for children's contribution of labor to the household declines, and as lower infant mortality means that survival to adulthood is more likely and thus fewer births are needed. To the extent that overpopulation contributes to environmental degradation, such human capital investments can lead directly to reduced stress on natural capital resources.

The Precautionary Polluter-Pays Principle

This plan is an elaboration on the polluter-pays principle mentioned earlier. To explain how it might be made into a policy instrument, we must take a brief look at the realm of certainty, risk, and uncertainty, and the differences between them.

Certainty is fairly simple. For example, "The sun will rise tomorrow" is about as close to a statement of true certainty as any mortal can make. Knowledge about some events is both widespread and sure.

As for *risk*, consider that roughly 40,000 people die on the highways every year in the United States. There is a finite likelihood (risk) that any given person will be one of them. One's actual risk depends on whether and how one drives, whether one drinks and drives, and so on, but it can be estimated statistically. Insurance companies do such calculations all the time to set their rates. We can say with confidence that there will be a certain number of highway fatalities in a given year and be accurate to within a fairly small margin, even though no one can say exactly who will be the victims.

Finally, there is true *uncertainty*. By uncertainty we do not mean blissful ignorance; in the case of natural capital, we know far too much about its indispensable functions now to ignore the signs of trouble. But there are a great many things we cannot be certain about. For example, it is truly uncertain whether global warming will happen; the average temperature rise that has been observed over the last 100 years could conceivably be due to natural climatic fluctuations. To many (probably most) atmospheric scientists, human-forced warming seems very likely, given what is known about the changing balance of greenhouse gases from human economic activity and our understanding of the relationship between concentrations of those gases in the atmosphere and global average temperatures. But all that can be said with assurance is that the globe seems to be warming and that current levels of several greenhouse gases are higher now than they were in preindustrial times. There is a wide range of estimates concerning how much warmer it might get, how much sea levels might rise, how weather patterns might change, and so on. But those estimates are broad and the probabilities quite tentative. Both are undergoing constant revision as more is learned.

Most environmental problems involve uncertainty. However, our social policy-making machinery — elected representatives and regulation-writing bureaucrats — understandably (if unreasonably) seek certainty, or, second best, clear estimates of risk. This leads to conflicts:

> Problems arise when regulators ask scientists for answers to unanswerable questions. For example, the law may mandate that the regulatory agency come up with safety standards for all known toxins when little or no information is available on the impacts of these chemicals. When trying to enforce the regulations after they are drafted, the problem of true uncertainty about the impacts remains. It is not possible to determine

with any certainty whether the local chemical company contributed to the death of some people in the vicinity of the toxic
waste dump. ...Because of uncertainty, environmental issues
can often be manipulated by political and economic interest
groups (Costanza, 1994, p. 399).

The precautionary polluter-pays principle (hereafter 4P) is one possible
response to this problem. It involves two ideas. The polluter-pays aspect
of 4P, as mentioned earlier, is simply to require that those who deplete
natural capital pay the full costs of doing so. The precautionary part refers
to the wisdom of trying to prevent ecological damage rather than blindly
going ahead and only afterward learning exactly what was lost and how
important it was. The alternative is to wait for science to deliver utter
certainty about the costs and benefits of doing something that would incur
ecological damage. But since science deals heavily in probabilities, revises
constantly, and raises several questions for every one it answers, such
certainty recedes from us, horizon-like, as fast as we approach it.

When scientific uncertainty is high (as it often is with the environment)
and especially when the ecological and/or economic stakes are also high,
radical politicization and conflict are often the result. Those who want to
proceed with a natural-capital-depleting economic enterprise can rightly
argue that there is no proof that it will cause serious harm. Those who
oppose the enterprise on the grounds that it might cause such harm, and
that no one knows for sure, can equally rightly argue that we should not
incur such risks without taking out insurance of some sort. So the trick
is to develop a policy instrument that addresses the problem of uncertainty,
requires economic actors to be responsible for the full and true costs of
their use of natural capital, and structures things so that the actors are
encouraged to reduce such costs.

One such instrument is the flexible environmental assurance bond. In
principle it is much like the deposit/refund system many states and
communities require for glass and plastic beverage containers, and would
be layered on top of some pay-as-you-go system (such as pollution fees)
designed to require polluters to pay for known damages. Each polluter
(say, a manufacturing firm) would be required to put up an assurance
bond equal to the amount of the best current estimate of the greatest
potential environmental damages that might occur in the future as a result
of the firm's activity. The money would be kept in an interest-bearing
escrow account for a predetermined period. At the end of that time, the
money would be returned (with interest) if the polluter could show that
no damage had occurred. In the more likely event that some damage had
occurred, sufficient funds would be withheld to repair the damage and

compensate injured parties. Firms thus would have an incentive to pollute less than the worst case. The system would also place the responsibility for ensuring proper regard for natural capital on those who extract and use it. The bonds could be administered by the Environmental Protection Agency or by an independent federal body created for the purpose.

Similar bonding systems already exist. Construction companies are often required to buy performance bonds, which may be forfeited if the work is not completed properly. This requirement helps prevent financially weak firms from winning contracts, but that is for the good, and would also be beneficial in the case of environmental assurance bonds. Competent, responsible firms that were too small to secure the required bonds by themselves could form co-insurance associations.

Ecological Tariffs

Tariffs have a bad name, primarily because they have often been used to shield inefficient and/or politically powerful domestic industries from the threat of cheaper imports.* A great deal of energy and diplomatic initiative has been expended over the years to lower or eliminate tariffs and other trade barriers and thus expand international trade, especially since the 1944 Bretton Woods conference created the International Monetary Fund, the International Bank for Reconstruction and Development (the World Bank), and the General Agreement on Tariffs and Trade (GATT).

The reasons why this is widely believed to be a good thing have to do with the economic theory of absolute and comparative advantage. Absolute advantage refers to the fact that countries differ in the quality of the factors that go into economic production, such as the richness of their natural capital stocks, the climate, the educational level of the workforce, the amount of capital available to invest in efficient means of production, and so on. If one country is better endowed with these factors than a competitor, its productivity will likely be higher and it will be able to make a given product more cheaply.

Comparative advantage, basically David Ricardo's insight, is a more subtle idea that is linked to the benefits of specialization. Ricardo illustrated the principle with a two-goods/two-countries model of the wine- and

---

* They have also been used to protect domestic social values, as expressed in laws designed to prevent child labor and other evils, by shielding domestic producers from cheaper competing products made by foreign companies unhindered by such legislation (see Schlefer, 1993).

cloth-producing capacities of England and Portugal (Daly and Cobb, 1989). Both countries have a finite supply of labor that can be apportioned between wine making and cloth production. Since any given laborer can only do one or the other, there is an opportunity cost to production; making wine means foregoing some production of cloth, and vice versa. In Ricardo's example, to make a given quantity of cloth in England requires 100 men working for a year, while a given quantity of wine takes 120 men working for a year. In Portugal the same amount of wine takes only 80 man-years and the cloth only 90 man-years. Portugal can make both wine and cloth more cheaply than England — that is, has an absolute advantage in both — so the question arises, why trade at all?

The answer is, because of the internal cost ratios. Portugal's cost of making a unit of wine is 0.88 units of cloth (80 ÷ 90 = 0.88). England's cost is 1.2 units of cloth. On the other hand, Portugal's cost of making a unit of cloth is 1.13 units of wine, while England's is only 0.83 units of wine. Portugal has a comparative advantage in wine and comes out ahead by specializing in wine production and then trading its wine with England in exchange for cloth. England (specializing in cloth) benefits too, because it exports cloth to buy wine more cheaply than it could make it at home.

This model has been theoretically expanded to encompass multiple goods and countries, and lies at the foundation of rhetoric in favor of expanded international trade. The central assertion is that by specializing according to their comparative advantages and trading among each other, all nations benefit because total production is thereby maximized and there is more wealth to share.

However, the matter is more complicated than that (see sidebar for additional discussion). Often ignored in the push for freer trade are Ricardo's assumptions. In particular, he assumed that labor and investment capital were immobile and confined within national boundaries. If this were not the case, then labor and capital would flow to the country with the absolute advantage in production. Today, however, both are considerably more mobile than in the early 19th century when Ricardo wrote. Capital is especially footloose: in these days of electronic finance, it can be shifted thousands of miles with a keystroke.

The doctrine of free international trade also assumes that there are no externalities involved in production. But environmental and social externalities are in fact commonplace, both locally and globally. Ecological tariffs may thus be necessary for the viability of graded ecozoning, the natural-capital depletion tax, the 4P assurance bonding instruments, or any other national policy designed to internalize environmental costs.

Here's why: Suppose one nation, Ecologicland, passes laws requiring firms to internalize their environmental externalities by paying taxes on their natural capital use and securing bonds to protect against unanticipated environmental damage. This is laudable, because those laws would be a step toward true efficiency, in that they would actually account for environmental costs. Suppose, however, that the neighboring Republic of Pollutia has no such laws. Other things being equal, if firms in Ecologicland act in the public interest, paying the taxes and securing the bonds, then their manufacturing costs will be higher than competing firms in Pollutia. Without tariffs levied by Ecologicland to raise the prices of imports, ecologically irresponsible Pollutian firms could flood Ecologicland with cheaper goods priced at levels that failed to reflect the environmental costs of their manufacture. This would put responsible firms in Ecologicland at a severe competitive disadvantage, possibly driving them out of business or encouraging them to relocate. If it happened widely, this process would have the effect of ratcheting down environmental protection standards worldwide.

Would tariffs used carefully in this way have destructive effects on the world economy? Not necessarily. Tariffs levied in the service of narrow-minded protectionism can reduce the efficiency of the global economy by thwarting the exercise of nations' comparative advantages. But tariffs used to prevent the undermining of a nation's efforts to internalize the costs of natural capital depletion would actually promote a more honest kind of efficiency. Moreover, the costs of tariffs may be exaggerated:

> Just how expensive is protectionism? The answer is a little embarassing, because standard estimates of the costs of protection are actually very low. America is a case in point. While much U.S. trade takes place with few obstacles, we have several major protectionist measures, restricting imports of autos, steel, and textiles in particular. The combined costs of these major restrictions to the U.S. economy, however, are usually estimated at less than three-quarters of 1 percent of U.S. national income. Most of this loss, furthermore, comes from the fact that the import restrictions, in effect, form foreign producers into cartels that charge higher prices to U.S. consumers. So most of the U.S. losses are matched by higher foreign profits. From the point of view of the world as a whole, the negative effects of U.S. import restrictions on efficiency are therefore much smaller — around one-quarter of 1 percent of U.S. GNP. ...Without a doubt, the major industrial nations suffer more...from unglamorous problems like avoiding traffic congestion and unnecessary

waste in defense contracting than they do from protectionism. ...[T]he cost to taxpayers of the savings and loan bailout alone will be at least five times as large as the annual cost to U.S. consumers of all U.S. import restrictions (Krugman, 1990, p.104).

Despite the ever-greater integration of the global economy, most nations continue to sell most of the goods and services they produce to themselves. That is why even a major trade war, in which sky-high tariffs cut international trade by half would cost the world economy only about 2.5% of its income (Krugman, 1990).

Natural Capital and International Trade

The wariness displayed by many environmentalists toward so-called international free trade is not just reformist ire or paranoia. Even mainstream economists have long recognized that free trade is not necessarily an unmixed blessing, that it helps some people and nations in some places and times, and hurts others. Paul Samuelson, a Nobel Prize winner and one of the deans of mainstream economics, put it this way more than 30 years ago: "What in the way of policy can we conclude from the fact that trade is a potential boon? ...We can actually conclude very little. ...Free trade will not necessarily maximise the real income or consumption and utility possibilities of any one country [nor will it] necessarily maximise the income, consumption and utility possibilities of a subset of persons or factors within a country" (cited in Ekins et al., 1994, p. 2).

The conventional argument in favor of free trade in general is that it makes for a bigger economic pie and thus there is more to go around. When it comes to the relation between trade and the environment, there is a related conventional argument: Trade promotes economic growth, which creates wealth, which, if fairly evenly spread around, creates more and more relatively well-off people who can afford tender attitudes toward the environment and its inhabitants. Trade therefore not only helps create the funds necessary to address the problems of environmental degradation and cleanup, it also helps the seeds of concern to sprout and make those funds available.

One flaw in this argument is that it is like stepping in quicksand: the harder you struggle, the deeper you sink. If existing economic activity generates both growth and environmental damage, more activity will lead to additional damage, requiring more growth to free up additional funds for damage mitigation, and so on.

Another concern is that even if resources are available to spend on repairing environmental damage, they must actually *be* spent that way, which is a different policy issue. Moreover, the damage done must actually be reparable. In many cases, such as those involving species extinction, it is not (Ekins et al., 1994).

As noted in the section on ecological tariffs, free trade is designed to further the cause of economic efficiency, which is economics' Holy

Grail. But efficiency is only real if all the costs of production and exchange have been honestly accounted for. Firms and nations competing against each other seek to reduce costs, and one way is to ignore them, i.e., to externalize them if possible by lowering standards. The temptation to do this has led to all kinds of historical and current abuses — child labor, dangerous worksites, etc. — which gave rise first to the labor movement and a great many laws protecting workers (Schlefer, 1993), and now to efforts to ensure that the environment is protected against nations or firms that compete by ignoring the costs of natural capital degradation. It is a see-saw battle; as currently written, for example, GATT allows nations to restrict imports of products shown to be produced with prison labor, but does not allow such restrictions in the case of child labor (Daly, 1993).

Free trade can stress natural capital by encouraging unhealthy patterns of commerce and resource exploitation. Many developing nations have specialized in one of the few areas where they have a comparative advantage: the harvest and export of primary products such as timber. A dangerous cycle often ensues: the output of primary products is boosted in an effort to increase foreign earnings, but the increase in supply drives prices down (real prices of non-oil primary products have fallen 45% since 1960). The option to make more money per unit output by processing the primary products and thus adding value to them is often foreclosed by tariffs levied at the export market end (e.g., the developed countries), which do not want the competition in intermediate and finished products. This pattern makes it difficult for developing countries to incubate other spe-

cialties, locking them in to a kind of subsistence-level "forced trade" (Ropke, 1994).

A more subtle effect of trade on natural capital can be seen in land use patterns in developing countries that attempt to export agricultural products. Land that is valuable for such crops is vulnerable to takeover by powerful economic interests. The displaced subsistence farmers then move on to less fertile lands, such as forests, where they clearcut, try to farm crops unsuited to the soils, and in other ways cause considerable environmental damage (Ekins et al., 1994).

One of the environmental externalities of trade is the effect of transport. Huge quantities of oil — perhaps one-eighth of world consumption — are devoted to moving products from one nation to another (Ekins et al., 1994). The natural-capital external costs of that consumption are not included in the prices of the shipped goods. Furthermore, these externalities "...have existed for decades, and an enormous number of decisions have been made under 'false assumptions.' We thus have a case of 'accumulated externalities,' where the distortions have been built into the physical and social structures of societies and into the corresponding trade patterns" (Ropke, 1994, p. 18). The more the volume of trade grows, the more pronounced become the externalities.

In sum, "...adherents to the hypothesis that growth and environment interact positively face a very heavy burden of proof" (Ropke, 1994, p. 16). The same goes for trade and the environment; the presumption that the relationship is benign should not be automatic.

Property Rights Regimes

So far in this chapter we have been discussing policy instruments that could be deployed without any changes in current property rights regimes governing natural capital. Property rights regimes are the formal arrangements people make to define and control their use of natural capital. They include private property, public property, common property, open access property, and various permutations of these four. Workable property rights regimes are necessary but not sufficient prerequisites to sustainability (Hanna, 1994). Where property rights regimes fail, the result is often environmental disaster.

A well-known metaphor for this failure is biologist Garrett Hardin's famous "tragedy of the commons" (Hardin, 1968). In Hardin's scenario, many herdsmen share a common pasture. Each herdsman has a strong incentive to add as many cattle to his herd as possible, since he derives all the benefits from the sale of the cattle thus raised, and the loss of the grazing resources his cattle consume is borne by all the other herdsmen. In terms of total utility, if adding a cow is worth one to the herdsman who owns it, the cost to each of the other herdsmen is only a fraction of one. With this incentive structure, it is only a matter of time before every herdsman, seeking to maximize his own utility, increases his herd to the point that the total grazing demand on the pasture destroys it — and probably the herds along with it.

It is a grim story, and much has been made of it in the literature. The tragedy of the (open) commons has been seized upon as an argument for private ownership of all resources. This may be because we are accustomed to thinking in terms of only two possibilities for the ownership of resources: private ownership or open access* (Hanna, 1994; Berkes and Farvar, 1989).

But open access resources, the context of Hardin's scenario, is only one of several categories of property rights regimes existing along a spectrum from completely private ownership (*res privatae*) to open access, or no ownership at all (*res nullius*). In between lie various combinations

---

* Why the preoccupation with the tragedy of the commons? Perhaps because of its sheer dramatic value: spectacular failures are more entertaining than quiet successes. Moreover, the Western cultural stress on individualism and competition has colored its science and led its scientists to focus on competition in nature and to long ignore the many examples of cooperation. When the subject was human beings, they were seen as humans struggling with nature rather than humans-in-nature, as we noted in Chapter 1. Under these circumstances, maybe it is no wonder common property systems were overlooked (Berkes, 1989).

of public ownership (*res publicae*), which is quite familiar to Westerners, and common property systems (*res communes*), which are not so familiar. Yet the common property regime can boast hundreds of readily available examples of successful, long-term, sustainable management of resources.

The distinction between open access resources and communally owned resources is crucial. Resource managers and policymakers who think Hardin's model defines the possibilities may be inclined to believe that the dangers of open access resources in an era of scarcity will demand rigid government controls or extensive privatization. This is consistent with the Western tendency to elevate the rights of individuals above everything else. But communal management offers a third alternative. Though often forgotten or neglected, communal management is far from unknown to Western cultures and has ancient roots in Anglo-Saxon common law and Roman law (Grima and Berkes, 1989). It is commonplace elsewhere in the world. Even in the United States, communal management of coastal lobster fisheries in Maine (Acheson, 1989) and of forest resources by Wisconsin's Menominee Indians (Hawken, 1993) show that the system can work in the midst of an aggressive market economy.

Communal management is one of several ways of classifying people's relationships to property (Grima and Berkes, 1989). The four major ones are:

> ***Open-access, commonly owned resources***. This is the arrangement that leads to the tragedy of the commons. There are no management institutions, or those that exist don't work very well, and the rights of access are not well-defined or enforced; anyone can take whatever he or she wants. If demand for the resource outruns supply, overuse and degradation result.

> ***Limited-access, commonly owned resources***. Limited access means that a system has been set up to define and control access to the resource. In general, there are three ways to do this: a local institutional arrangement (communal property or management), management by government, or a combination of the two.

> ***Limited-access, privately owned resources***. In this case, private ownership confers clear and well-defined rights to exclusive use and to sell or otherwise transfer those rights. Non-owners can be legally refused access.

> ***Privately owned, but effectively open-access, resources***. Although resources may be privately held and intended for

authorized, exclusive use only, the owners may fail to achieve that condition. A large private game park, for example, may not be defensible against poachers.

In addition to these common possibilities, resources can also (though rarely) be closed-access and communally owned, as in the case of monasteries or the estates of utopian movements.

To Westerners accustomed to private ownership of everything of importance to them in their daily lives, communal ownership and management of important resources may seem strange, unworkable, or even contrary to human nature. Yet many cultures worldwide, both ancient and modern, have regarded the idea of private ownership of resources as inappropriate and have seen their responsibility as one of stewardship toward natural resources, of which they have been granted temporary use.

Common property systems confer certain benefits upon the communities that employ them (Berkes and Farvar, 1989):

- They help secure the livelihoods of the members of the community by ensuring that they will always have a way of making a living.

- They provide a means, through the mutually agreed-upon rules for managing the resource, for ensuring that everyone has fair access to the resource, and for resolving disputes over access.

- They can form an organizational basis of production.

- They are basically resource-conservative, and often aim at self-sufficiency. The associated value systems tend to emphasize taking only what is needed and to discourage excessive individual gain.

- Their resource conservatism and their tendency to use rituals to keep resource use in synchrony with natural cycles help to promote sustainable use over time.

Common property systems tend to be concentrated among traditional societies, and have often suffered in the face of the development of larger, market economies. The environmental ravagement seen in many developing countries can be attributed partly to the corruption or collapse of traditional common property systems of resource management (Goodland et al., 1989). The breakdown becomes a strand in a web of sweeping changes in the circumstances of resource use. For example, widening participation in market economies tends to encourage overexploitation of resources for export rather than extraction for local subsistence only. The increasing centralization of power, displacement of local control, and application of inappropriate pricing, subsidies, legislation, or other government incentives can destroy local communal property management arrangements. These policies may include building roads into forested areas to encourage heavier settlement, tax concessions to individuals or firms engaged in natural resource exploitation, subsidized agricultural credit that encourages over-use of fertilizers and pesticides, and land-titling policies that require "homesteading" (clearing) a plot of land before title is granted.

Where common property systems are subject to open access pressures, sequential exploitation of resources — from the most valuable and accessible to the least — takes place and frequently leads to the systematic depletion of resource stocks until such activity is no longer profitable (Grima and Berkes, 1989). A typical sequence of events might go like this: Communal management of given resources exists stably for some time, perhaps decades or even centuries. Various pressures of the sort described previously lead to free access to the resource. The existing common property systems and their allocation mechanisms collapse or are pushed aside, leading to heavy resource exploitation and sometimes ecological collapse of the resource.

Three fixes are possible. First, under the market solution, the resource can be privatized (provided it has some value left or can be restored). This can work if the resource-use rights are made exclusive and transferable. Such rights are relatively easy to define and enforce in some cases (agricultural lands), less so in others (fish and wildlife). Second, the central government can impose regulation. Third, a common property system can be revived, as with the Chipko movement in India (Berkes, 1989). In precolonial India, local communities managed large tracts of communally held and natural forest sustainably. Under British rule, the commons forests were turned into private property and the natural forests logged for commercial gain. In modern times, the Chipko ("to embrace" in Hindi) movement inspired villagers to embrace trees to prevent them from being cut down by loggers. It evolved into "a lobby group demanding ecological rehabilitation as well as social justice. A Chipko slogan says that 'soil, water and oxygen' are the main products of forests, emphasizing the importance of restoring the forest cover of the Himalayas. ...From a common-property point of view, the Chipko movement is an example of the reassertion of traditional communal resource-use rights" (Berkes, 1989, pp. 237-8).

Given the extent of plans for economic growth and modernization in developing nations, the pressures bearing on common property systems are heavy. But their resilience and value as tools for the long-term sustainable management of resources should not be lightly discarded. Widespread, abrupt, and heedless change can destroy such systems, but change at a pace that can be accommodated seems to make for the possibility of their survival and evolution.

Resource Utilities

Entrepreneur and writer Paul Hawken has proposed another approach to the management of natural capital for sustainability: the resource utility (Hawken, 1993). In some ways, the concept of a resource utility echoes some of the features of the common property arrangements discussed above. Utilities combine features of public and private institutions in that they are profit-making organizations that accept public control (in the form of regulation by public utility commissions) in exchange for monopoly access to a resource and a specified rate of return. In theory, they can be structured so as to maximize return from the resource over the long term, which means they can ignore the usual temptations to over-exploit the resource. They would have little inclination to externalize as many costs as possible, which is one of the major pressures on businesses in competition. Further, utilities' guaranteed profit reduces the risks for investors, thus allowing them to attract investment capital at relatively low interest rates. This helps make it easier to develop and carry out long-range plans.

Hawken uses a hypothetical salmon utility to illustrate the application of the public utility concept to renewable resources:

> A salmon utility would recognize that existing market mechanisms do not operate in the best interest of the fish, the fishermen, the consuming public, or the salmon habitat. To support the utility, there would be a fee on salmon landed on the Pacific Coast. Those revenues would go directly to a central Salmon Utility or a number of smaller regional salmon agencies, whose sole purpose would be to increase the stock of salmon. To do this, the utility would spend its funds primarily on habitat restoration, but also on education, land acquisition...and research. As a utility, it would be allowed a guaranteed profit of 10 to 12 percent.... The salmon utility would issue stock just as a power utility does, but given its guaranteed revenue from the salmon tax, it would also have the capacity to issue bonds at favorable interest rates, which could be used to invest in

> long-term restoration projects. ...The long-term result...would
> be the increase of wild salmon. This increase would in turn
> bring in higher revenues that would give the utility even greater
> capacity to carry out its agenda (Hawken, 1993, p. 193).

The potential for relatively enlightened resource management by utilities can be seen in the electric utilities that have operated demand-side
management (DSM) programs to promote energy conservation. The principle behind DSM programs is that utilities recast themselves as suppliers
of electricity services (e.g., lighting) rather than commodity electricity.
When in the role of traditional commodity suppliers, utilities operate on
the same principle as other businesses: higher sales mean more revenues
(as well as greater throughput and degradation of natural capital). But as
service suppliers, utilities can work to save customers money by encouraging (and even subsidizing) the use of high-efficiency equipment such
as compact fluorescent light bulbs and variable speed motors. By thus
reducing electricity demand growth, utilities can delay or avoid the enormous and risky investments in new power plants. They split the money
saved with their customers.

Like many of the proposals discussed in this chapter, the workability
of resource utilities depends less on the details of the plan than on the
strength and depth of the political will behind them. Traditionally, utilities
have been subject to regulation by state public utility commissions (PUCs).
The makeup and activities of PUCs remain, at best, vague to most citizens
and the commissions vary in the degree of their responsiveness to the
public mood. Since PUCs have traditionally defined the terms by which
utilities can make money, the relationship between a utility and its state
commission, and the commission's view of its charge from the public, are
critical to the utility's ability to carry out any plan, whether it is to spend
several billion dollars on a new nuclear power plant or invest in aggressive
DSM programs. The DSM activity undertaken by utilities in the last few
years would never have happened unless the governing PUCs had decided
to restructure their regulations so that DSM programs made economic
sense for utilities.

But utility commissions have always been subject to many pressures
besides those stemming from environmental concerns, and now the conditions under which the electricity industry operates are changing radically.
The guiding and restraining hand of regulation is relaxing its grip and the
result, among other things, is declines in utility spending on DSM. Early
in 1994, the California Public Service Commission sent shock waves
through the industry by announcing a proposal to phase in "retail wheel-
ing," which would eventually allow all electricity customers, from the

biggest factories to individual homeowners, to shop around for the best deal they could find on electricity. This proposal launched a movement toward more open competition that is transforming the sheltered world of regulated utility monopolies. Large users of electricity, such as industrial firms, support open competition because energy costs are a substantial fraction of their production costs. Many utilities, on the other hand, are deeply worried. Having made PUC-approved past investments in expensive power plants (often nuclear reactors), they fear that they will be unable to pay off the loans for those facilities if competition drives electricity rates down below their costs of generation.* More to the point of this book, many conservationists argue that it is the regulated monopoly arrangement that has allowed utilities to begin thinking of themselves as suppliers of services rather than commodities. Open competition, they believe, will be the death of conservation efforts because the regulatory protections that allowed utilities to earn income while selling less electricity will be gone. Lower rates will also reduce customers' incentives to conserve. By dismantling the incentive structure for conservation (lower throughput), open competition seems likely, at best, to force the system back to the uncertainty of reliance on command-and-control measures implemented by state governments and PUCs to reduce the natural-capital impacts of energy use. As things stand, free-wheeling competition among electricity generators (including non-utility generators) is driving them to use more cheap natural gas and coal, both carbon-based fuels (the latter the most carbon-rich type). Renewable sources, just recently beginning to approach competitive generating costs, are being pushed to the margins again. From being structured so as to encourage acceleration toward conservation, the system is being switched back into the retrograde mode of having to apply the regulatory brakes to discourage excessive exploitation of natural capital.

The point is that, while it is possible to imagine systems that would work better to reduce throughput and limit scale than the ones presently in use, the absence of popular understanding and support will render the most ingenious reform plan irrelevant.

---

* U.S. electricity prices have been falling since 1982 and are expected to decline by 1% per year through 2020, due to increased competition. As of September 1, 1998, 18 states had enacted legislation or issued comprehensive regulatory orders designed to promote competition. Most of the remaining states were considering such legislation or orders. See U.S. Energy Information Administration, *Annual Energy Outlook 1998*, p. 52, and http://www.eia.doe.gov/cneaf/electricity/chg_str/tab5rev.html.

Two Sustainability Profiles

The Netherlands

The Netherlands is a tightly packed nation of about 15.7 million people and a population density of about 4.6 people per hectare, the highest in Europe and among the five or six highest in the world.* Its economy is based heavily on industry (metals processing and chemicals), intensive agriculture (The Netherlands is second only to the United States in agricultural exports worldwide) and transshipping. The country is plagued by serious air pollution from industry and automobile use; water pollution from agricultural runoff, heavy metals, and organic compounds; and nitrate contamination of groundwater. Prompted by these severe environmental stresses, The Netherlands has become one of the most vigorous actors on the sustainability stage.

The heart of The Netherlands' effort is the National Environmental Policy Plan (NEPP), which was first issued in 1989 and aims officially to achieve sustainable development within one generation. Now in its third iteration, NEPP deemphasizes strict reliance on command-and-control mechanisms and end-of-pipe cleanup in favor of reshaping social and economic structures over the long term to focus on sources rather than effects, according to published materials from The Netherlands Ministry of Housing, Spatial Planning and the Environment. The plan also attempts to integrate solutions across environmental media (air, water, and soil).

NEPP organizes Dutch environmental policy according to nine themes: climate change, acidification, eutrophication (nutrient buildup in surface waters), toxic and hazardous pollutants, contaminated land (presence of xenobiotic pollutants in soil), waste disposal, disturbance (noises, odors, local air pollution), groundwater depletion, and resource dissipation (avoiding inefficient use of natural resources). Within each theme, each of several target groups — consumers, agriculture, industry, refineries, energy companies, retail trade, transport, construction firms, waste disposal companies, and firms involved in water supply and sewage treatment — is assessed responsibility according to its contribution to a problem. Agriculture, for instance, is held responsible for 64% of the eutrophication problem, 36% of acidification, 12% of climate change effects, and 3% of the waste disposal load.

---

* By comparison, the United States has a population density of about 0.3 people per hectare.

NEPP 2 set environmental quality objectives for each area and defined the emissions reductions required to meet them, generally by 2010. For example, the government determined that achieving a sustainable level of acidification would require reducing emissions of sulfur dioxide and nitrogen oxides to 20 to 30% of their 1985 levels. The plan likewise aimed to reduce emissions of eutrophic chemicals such as phosphates and nitrates by 70 to 90%; to reduce emissions of "priority substances" (heavy metals and pesticides) by 50 to 70%; to shrink the waste stream by 70 to 90% and reduce landfilling of waste by 80 to 90%; and, more modestly, to stabilize carbon dioxide emissions at average 1990 levels by 1995 and reduce them by 3 to 5% by 2000.

NEPP 2 and 3 spell out a mix of regulations, incentives, and voluntary agreements to achieve its ends. The Ministry of the Environment is overall coordinator but shares responsibility for NEPP's implementation and development with the ministries of agriculture, transport, economic affairs, finance, and foreign affairs. Target groups representing the sources of environmental problems are courted via an open planning and negotiation process, which has led to the drawing up of sector plans with all target groups and to the conclusion of a number of voluntary agreements (covenants) with industry that spell out voluntary emissions reductions or broader environmental action plans. The covenants include requirements for industry reporting of emissions. Financial incentives include local charges for waste, sewerage, and water pollution; environmental taxes applied to fossil fuels, groundwater, and waste; and consumer incentives such as product deposit/refund systems, tax incentives for clean products, and high gasoline taxes. Social support for NEPP's aims is encouraged by means of environmental education in schools, public education campaigns, and subsidies for environmental organizations.

In documents describing NEPP published in early 1998, the government claims to have achieved an "absolute decoupling" of economic growth from increased strain on the environment and, except for greenhouse gas emissions, points to reduced pollution as evidence. Costs of environmental policies have increased over the years, reaching 2.7% of GDP in 1995. They are expected to settle at around 2.5% of GDP per year by 2010.

NEPP 2 was introduced in 1993 and ran through 1998; NEPP 3 is in effect from 1999 through 2002. New and adapted target-group policies include the following:

- ***Consumers and the public.*** Continue to encourage modification of individual behavior through education and information and by working to ensure that prices for goods reflect their environmental costs.

- **Agriculture.** Focus on animal waste, including new legislation to restructure the pig-farming industry. Seek a covenant with the fertilizer industry to ease groundwater pollution.

- **Industry.** Encourage continuous improvement in environmental performance; postpone deadline for achieving industry nitrogen oxides emission targets from 2000 to 2005 but threaten regulation if industry does not conceive means of achieving targets by then; develop product stewardship systems to reduce waste.

- **Refineries.** Tighten emission standards for sulfur dioxide and nitrogen oxides to reflect the potential of developments in new technologies.

- **Energy companies.** Though clearly a major sector, it is not clear just what new initiatives are in store for energy; the government says that it is pursuing a "joint approach" with refineries and the industrial sector to reduce emissions. NEPP 3 notes that although carbon dioxide and nitrogen oxide emissions are growing more slowly than energy consumption, a major effort will be needed to achieve the targets set in earlier versions of NEPP. Nitrogen oxides targets, in fact, are not considered achieveable without incurring sharp increases in mitigation costs.

- **Retail trade.** The retail trade sector has agreed to improve its energy consumption by 23% by 2000 (1989 base); retailers are expected to stock environmentally friendly goods and be active in using environmental management systems.

- **Transport.** The transportation sector is implicated in the problems of climate change, acidification, toxic and hazardous substances, and noise pollution. Emissions of carbon dioxide are increasing, those of nitrogen oxides and noise are decreasing. However, year-2000 targets for all three will probably be missed due to strong growth in traffic volume. The government aims to limit road transport growth, improve public transport, encourage new pollution-control technologies, influence driving and vehicle-purchasing behavior, and make transportation infrastructure use more efficient.

- **Waste disposal.** The government claims that major reductions were achieved between 1990 and 1995 in incinerator emissions of important pollutants, including dioxins and heavy metals. New taxes on the landfilling of combustible waste are designed to make that option equal in cost to incineration. Internal barriers to transport of waste are being eliminated and the government seeks the same goal for international barriers, presumably thus allowing The Netherlands to export waste.

Costa Rica

Costa Rica, with a population of 3.7 million, is one-fifth the size of The Netherlands and offers several other contrasts as well (see Table 6.1). Its resource base is marked by a lack of minerals and an abundance of hydropower potential. Its tropical habitats house an immense store of biodiversity, which the nation seems determined to conserve and develop sustainably. As noted in Chapter 4, Costa Rica's major environmental problems include extensive deforestation and related soil erosion.

Costa Rica's sustainability goals are perhaps less explicit than those of The Netherlands but seem equally ambitious. Led by visionary President Jose Maria Figueres, the government in 1994 officially declared its intent to restructure the economy for sustainability and developed a related 10-point general action plan. Even given the vagaries of politics and the advisability of regarding the plan as a "central tendency with a cloud of variance" (Janzen, 1998) rather than commandments chiseled in stone, most of its points seem to have survived the transition to a new, more conservative government in 1998, although with increased emphasis on privatization. The plan is summarized below from material in statements by government officials and published government documents:

- To promote adequate valuation of environmental resources, including internalizing environmental costs in prices of goods that require large inputs of natural resources or have large environmental impacts. An example of this is a 1998 agreement between the government and Grupo del Oro, an orange juice company with a facility near the Guanacaste Conservation Area. Del Oro will pay the government nearly half a million dollars over 20 years in recognition of the ecosystem services provided by the 1,200-hectare conservation area, including biological pest control, biodegradation of orange peels, water supply, carbon fixation, and others (Janzen, 1998).

- To design modern and appropriate management techniques for the National Conservation Area System, including emphasis on

Table 6.1 Measures of development and sustainability in the Netherlands and Costa Rica

	The Netherlands	Costa Rica
Population, 1998	15,739,000	3,650,000
Average annual pop. change, 1995-2000 (%)	+0.5	+2.1
Total fertility rate, 1995-2000	1.6	3.0
Life expectancy at birth, 1995-2000 (years)	77.9	76.8
Per-capita gross domestic product,		
purchasing power parity basis, 1995 (Int$)	19,880	5,920
Income distribution, % per population quintile		
0-20	6.9	4.0
20-40	14.2	9.1
40-60	18.9	14.3
60-80	23.7	21.9
80-100	36.4	50.7
Spending on public education, % of GNP	5.3	4.5
Carbon dioxide emissions, 1995 (metric tons)	135,909,000	5,232,000
global share (%)	0.60	0.02
per-capita	8.8	1.5
Protected areas, 1997 (hectares)	228,000	702,000
% of total area	6.7	13.7
Biodiversity		
total, known mammal species	55	205
threatened " "	6	14
total, known breeding bird species	191	600
threatened " " "	3	13
total, known higher plant species	1,170	11,000
threatened " " "	1	456
total, known reptile species	7	214
threatened " "	0	7
total, known amphibian species	16	162
threatened " "	0	1
Total forest (including plantations), 1995 (hectares)	334,000	1,248,000
% change, 1990-1995	0.0	-3.1

Source: *World Resources 1998-99*, World Resources Institute, Washington, D.C.

community participation and coordination with tourism. The goal of the system is to ensure the long-term preservation of the one-fourth of the country that is currently earmarked for conservation and to expand that territory until it comprises one-third of the country's total area. The aim of the current administration is to create effective incentives for private landowners to preserve forest habitat rather than to create new public protection areas. The government has begun a biodiversity inventory and aims to classify every species in the especially species-rich Guanacaste Conservation Area within five years. Other conservation areas are also being surveyed.

- To adopt and promote sustainable production technologies in agriculture, tourism, and manufacturing. One planned means is the use of environmental certification programs (green labeling). The government is seeking investment in sustainable production from private investors, other national governments, and international lending institutions, and has created the National Fund for Sustainable Development as a vehicle. The Fund issues Environmental Certificates, financial instruments that can be used, for example, for debt-for-carbon swaps or for funding biodiversity research. In agriculture, the policy is to emphasize products such as lean, "green" beef, and fruits (such as oranges and bananas) grown without the use of chemical pesticides. The government also aims to recover abandoned pastureland and achieve an integrated agroscape employing various but standardized farm plans for the varying topographies.

- To step up control of water, air, and soil pollution, including waste cleanups and restoration of damaged waters. Lead was phased out of motor gasoline in 1995.

- To develop alternative energy sources and to encourage reforestation to preserve hydroelectric potential. Costa Rica's electric power companies are working to eliminate coal-fired generation in favor of hydropower (which already generates most of the country's electricity), wind, geothermal, biomass, and waste-to-energy generation. The government expects that 98% of all electric power will be generated by these means by 2000 and plans to eliminate the importation of fossil fuels for electricity generation by then. In 1994 the Global Environment Facility sponsored a $3 million windpower project in Costa Rica. A proposed new law would eliminate limits on private renewable generation.

- To promote education about and community participation in sustainability projects and activities, particularly those related to the country's biological diversity. These efforts are to be integrated with promotion of ecological tourism and regulations governing tourism that reduce its environmental impact. A related effort that helps address the goals of both community participation and biodiversity preservation is the parataxonomist program run by INBio, the National Institute of Biodiversity. The program trains and employs rural residents to gather and categorize biological samples.

- To seek out and take part in appropriate international agreements. Costa Rica has ratified the Conventions on Biological Diversity and Climate Change that emerged from the 1992 United Nations Conference on Environment and Development. In 1994, Costa Rica signed a letter of intent with the United States to support bilateral sustainable development, cooperation, and joint implementation of ventures to mitigate the effects of climate change, among other sustainability efforts. INBio has signed several "bioprospecting" contracts with major pharmaceutical companies to promote development of pharmaceuticals derived from or based on native plants and animals.

- To stem uncontrolled development and promote the ordered distribution of people, economic activity, and conservation areas throughout the country by means of land-use planning, zoning, etc.

- To strengthen and reengineer institutions to ensure inclusion of all sectors — government, industry, academe, the public at large — in the creation and coordination of sustainability policy and action. A step toward this goal is the creation of the National Sustainable Development System, composed of the National Sustainable Development Board, the Foundation for Sustainable Development in Costa Rica, the Technical Advisory Commissions, and the Sustainable Development Area Board.

- To ensure that government sets a good example by adopting policies to save energy, recycle, and manage waste responsibly.

Afterword

...."Hurry up! We have to change all the packages!" the three section chiefs said. "The Society for the Implementation of Christmas Consumption has launched a campaign to push the Destructive Gift!" "On the spur of the moment like this," one of the men remarked. "They might have thought of it sooner...." "It was a sudden inspiration the president had," another chief explained. "It seems this little boy was given some ultramodern gift articles, Japanese, I believe, and for the first time the child was obviously enjoying himself...." "The important thing," the third added, "is that the Destructive Gift serves to destroy articles of every sort: just what's needed to speed up the pace of consumption and give the market a boost.... All in minimum time and within a child's capacities.... The president of the society sees a whole new horizon opening out. He's in seventh heaven, he's so enthusiastic...." "But this child..." Marcovaldo asked, in a faint voice: "did he really destroy much stuff?" "It's hard to make an estimate, even a hazy one, because the house was burned down...."

— Italo Calvino
Santa's Children

151

Our culture holds several beliefs that are probably unsustainable. To one degree or another, most of us accept that fulfillment lies in acquiring things, that perpetual economic growth is both normal and the answer to all social problems, that other species and natural capital in general are simply ours for the taking rather than elements of a shared ecology, and that technology will save us from the need to make moral choices about wealth and poverty. This book has advanced a point of view and some policy options that are incompatible with these beliefs and that would be reinforced by a different body of values and related behaviors.

There are two arguments for changing our policies, values, and behaviors with respect to natural capital. The first is practical. This book takes an instrumental viewpoint: we argue explicitly that we must be more careful with natural capital because we need it to live. Instrumentalism asks, "How can we use natural capital and what is the price?" We have not thus far believed it necessary to be very careful in using or pricing natural capital. One sign of this is the notion that further, even endless, economic growth — specifically, increasing the total scale of the global economy — is not only possible and desirable, but is a prerequisite for environmental health. Only when people have surplus wealth, the argument goes, will they be willing to pay the costs of cleanup, pollution control, and other environmental programs.

It is true that some features and services of natural capital probably are discretionary from a strictly instrumental viewpoint. If all the endangered species in North America were to disappear tomorrow, it is doubtful that ecosystem services would suffer noticeably. The contribution of most of those species is already marginal. Nevertheless, the aggregate flows and services of natural capital in general are a prerequisite to economic activity. The trouble with the growth-leads-to-ecohealth argument is that it is self-contradictory (the growth that creates the cleanup fund also creates more pollution, requiring more wealth and thus more growth, etc.) and that it casts natural capital as a discretionary luxury. It suggests that we can take care of the environment once we've "got ours." We argue that in fact there will eventually be nothing for us to get unless we tend more carefully to the Earth's endowment of natural capital. A more rigorous instrumental attitude toward natural capital would force us to appreciate its indispensability and thus do us, and it, good.

The second argument has to do with living meaningfully. The emptiness of much of modern life, especially its materialism, has become a cliche. Those who feel this intuitively but wonder if they are out of step with the rest of the world might be intrigued by a remarkable and now classic study by economist Richard Easterlin (1974). Easterlin gathered and analyzed surveys of self-reported human happiness in 19 rich and poor

countries since World War II. Not surprisingly, the things of most importance to personal happiness were economic well-being, family, and health.
What was striking, however, was a pattern now known as the Easterlin
paradox:

> Within countries there is a noticeable positive association
> between income and happiness — in every single survey, those
> in the highest status group were happier, on the average, than
> those in the lowest status group. However, whether any such
> positive association exists among countries at a given time is
> uncertain. Certainly, the happiness differences between rich and
> poor countries that one might expect on the basis of within-
> country differences by economic status are not borne out by
> the international data. Similarly, in the one national time series
> studied, that for the United States since 1946, higher income
> was not systematically accompanied by greater happiness (p.
> 118).

In other words, rich Americans were happier than poor Americans,
and rich Indians were happier than poor Indians — but Americans on
average were no happier than Indians on average, though they averaged
greater wealth. The reason is that people compare themselves to the
Joneses next door. Happiness is judged on the basis of relative wealth
and the norm for one's own society, and most Americans have no personal
experience with the Joneses in India. Americans judge themselves happy
to the extent that they perceive themselves to be ahead of, or at least
keeping up with, the American Joneses.

Our ignorance of this effect chains us to life on a treadmill. Easterlin
bluntly concluded that "[o]ver time...as economic conditions advance, so
too does the social norm, since this is formed by the changing economic
socialization experience of people. ...*[E]conomic growth does not raise a
society to some ultimate state of plenty. Rather, the economic growth process
itself engenders ever-growing wants that lead it ever onward*" (p. 121;
emphasis added).

If happiness, as we now define it, always depends on having more
than the people next door, and if only perpetual economic growth can
satisfy this condition, then only such growth can keep us happy. Yet, as
this book has tried to point out, there are good reasons to believe that it
is unsustainable.

This suggests trouble ahead. We like our creature comforts and are
inclined to surliness when they are threatened. As this unease widens,
say, in times of recession, it tends to express itself politically in civil

distress. This is easy to see in places that seem perennially troubled and close to the economic margin. In famine-wracked Somalia, those with guns eat well. But it can happen anywhere, even in rich societies. In late 1994, the United States held midterm elections that, reflecting the voters' mood, were remarkable for their venom and divisiveness. Few Americans starve, but the slippage of living standards and the widening gap between rich and poor fuels an unfocused anger. And as archaeologist Joseph Tainter notes in *The Collapse of Complex Societies*, the survival of political structures can depend on the continued creation of wealth:

> ...[S]upport for leadership must...have a genuine material basis. Easton suggests that legitimacy declines mainly under conditions of what he calls "output failure." Output failure occurs where authorities are unable to meet the demands of the support population, or do not take anticipatory actions to counter adversities. ...Output expectations are continuous, and impose on leadership a never-ending need to mobilize resources to maintain support. The attainment and perpetuation of legitimacy thus require more than the manipulation of ideological symbols. They require the assessment and commitment of real resources, at satisfactory levels, and are a genuine cost that any complex society must bear (p. 28).

The point is that our way of life and our primary source of meaning, defined both personally and politically, are dangerously dependent on a steady flow of wealth (manufactured capital), just as we seem to be entering an era when restraint is necessary. In our personal lives, we work to acquire the next house, the next car. Our political system aims (at least in theory) to maximize economic efficiency, on the grounds that efficiency maximizes wealth. Yet, because of the constraints imposed by natural capital limits, stepping off the treadmill, or at least slowing it down, may be the only way to ensure ourselves and future generations a fair chance at personal happiness and fulfillment as well as the survival and vitality of the political structures that make community possible.

What does this require? The authors believe that the outlook and policies discussed in this book would be a long stride in the right direction. These policies — green taxes, graded ecozoning, the precautionary polluter pays principle, and so on — would work in concert with citizen activism and personal environmental responsibility to transform the economic system so that its automatic tendency, its "default mode," would be to conserve and invest in natural capital. The policies would need a base of popular support to be implemented, and as they were implemented they would restructure the system in ways that would reinforce the values that supported them.

It is also likely that this transformation would help revive our sense of community, i.e., of an intimate connection to the places where we live and the others we share them with. Many of us say, regretfully, that we lack a sense of community. Yet our current economic system and its assumptions tend to thwart community by treating people as economic atoms, encouraging our indebtedness to distant sources of supply and waste disposal, and obscuring our ties to the ultimate (natural) sources of our wealth. The changes proposed here would alter the system in ways that would undermine these circumstances and thus encourage and support community.

An economic system altered as we propose probably implies a stronger focus on our local environments. The Internet and other forms of electronic interconnection make it possible to talk with people across the globe about shared interests. But the Internet is faceless and there are no places in cyberspace. True community, to the extent that it entails not only social connectivity but also close links to place, requires locality. The social costs of loss of community are high; so are the ecological costs of the loss of locality, of our sense of where we live. No longer tied to the land, we are estranged from the sources of our wealth, and estrangement leads to abuse. In Chapter 2 we saw that cities, where most of us live, are on ecological welfare. Every city's ecological footprint extends far beyond the city limits and most of our wealth comes from far away. Although we know that our water, for example, doesn't really come from a tap, or pork chops and vegetables from the grocery store, few of us really understand where they do come from. Was that water pumped from a well or drawn from the river? Was that hog raised in the next county or the next time zone? Are those grapes from California or Chile? Ignorant of the origins of our food, we cannot say how it was grown, how the land was treated, how much energy was used to produce it and get it to our tables, or what social arrangements — the labor of machines or perhaps of children — made it available.

Efforts toward sustainability and community can reinforce each other, according to four principles (Ropke, 1994):

1. Wants are largely formed by the social and physical structures of everyday life. Those structures can be changed to reduce material wants, as when a community decides to emphasize public transit rather than build more roads.

2. Achieving local control means economic decentralization and a greater degree of local self-sufficiency. Members of the community depend more on each other and less on others far away. More

decisions about the economic fate of the community can be made by those directly affected.

3. Local efforts are most effective at tightening or closing the circuits of economic production to address environmental problems.

4. The closeness of local economic relationships (and their consequences) reinforces the ethical bonds and constraints of community.

The local focus is also the only antidote to the danger of abstracting one's love of place, of loving Yosemite from afar and then pouring the used crankcase oil down the storm drain. It may be easier to love some places than others, but as Wes Jackson of the Land Institute in Salina, Kansas, put it, "It is possible to love a small acreage in Kansas as much as John Muir loved the entire Sierra Nevada. ...Either all the earth is holy, or it is not" (1991, p. 51).

Correcting the failures of the current economic system does not necessarily mean sweeping away the entire existing order. We are not going to stop being an urban species, nor give up civilization and its comforts and enrichments. We are not all going to return to the land and live as subsistence farmers. But the restructuring we propose would encourage the cultivation of a more intimate awareness of and attachment to the places where we build our dwellings and make our homes and livings. It would encourage and support a culture of environmental stewardship emotionally and economically uncongenial to reckless overconsumption. That would go a long way toward addressing our natural capital survival problem.

We (the authors) see the cultivation of ecological economic values and the introduction of the policies for natural capital management discussed in this book as mutually reinforcing processes. The emerging ecological values would create popular support for the policies, and the policies would adjust the economic system so that market forces would help people and corporations to act sustainably rather than destructively. We conclude with a discussion of five general ecological economic values and the ways in which they and the policy proposals interact with and support one another:

1. ***The value of knowing our ecological place.*** If we take the time and effort to find out where our food, energy, and water come from, to read the labels on our clothing, to learn where our automobiles or bicycles were built, and so on, then we can begin to see how distant are the sources of everyday things. The vastness of the economic web we live in isolates us from the natural-capital consequences of our purchases. Wendell Berry has written that

"we have allowed our suppliers to enlarge our economic boundaries so far that we cannot be responsible for our effects on the
world. The only remedy for this...is to draw in our economic
boundaries and shorten our supply lines so as to permit us literally
to know where we are economically. The closer we live to the
ground that we live from, the more we will know about our
economic life [and] the more we will be able to take responsibility
for it" (1992, p. 35).

Our supply lines are long for a number of reasons. For example,
ignoring the environmental costs of crude-oil consumption causes
its underpricing and thus the underpricing of transport. Imposing
natural capital depletion taxes to reflect some of those environmental costs would raise oil prices, make transport more costly,
and encourage greater reliance on local sources of the transported
items. Another reason is that unconstrained international trade
allows nations with low environmental standards to produce items
at lower cost than nations with higher standards, and then export
them to the latter. This practice discourages ecologically responsible
national suppliers and impels consumers to buy from foreign
sources. Careful use of ecotariffs would level the field, allow
national suppliers to compete with irresponsible foreign suppliers,
and prevent consumers from being penalized for ecologically
responsible purchases.

2. *The value of thinking in circles.* Earthly natural processes
generally operate in the round, yet we usually think of them in
straight lines, as when we buy something from an unknown origin,
use it up, and throw it "away." Objects seem to enter our lives,
pass through them and disappear on more or less linear trajectories.
In fact, of course, "away" is not the end of the line, and the
trajectories are actually curved. The importance of this fact — what
goes around, comes around — is that it is another way of saying
that the Earth is a closed system and that what happens here, and
its consequences, stay here. Those that persist or exceed the
Earth's capacity for amelioration must eventually be dealt with, if
not by us, then by our children or their children.

Once again, getting the prices right would reinforce this way
of thinking and help guide us toward sustainable behavior. For
example, few products incorporate into their prices the costs of
their disposal or the externalities imposed on the environment or
other people during their manufacture. Proper pricing would
encourage intelligent design for recycling or reuse and encourage

us as consumers to hold out for more durable, well-made, and well-designed products.

3. *The value of scaling back.* The increasing scale of the global economy — ever-increasing numbers of people all wanting more — is one of the gravest threats to natural capital and long-term human economic survival. Control of scale means control of both population and consumption. Overconsumption is the developed world's defect, and it is to people living in the developed world that the suggestion to scale back primarily applies.

Scaling back means living with a lower rate of throughput. At the personal level, it means spending less and saving more, and perhaps working less and trading lower incomes for greater leisure. It doesn't mean subsistence, but an attitude of greater reverence for things and the realization that to gain them is to lose natural capital. Paradoxically, by making things so easily acquired, industrial society invites disdain for them. A toaster is a minor miracle, but when one fails, typically it's just thrown away rather than repaired. The way the economic system is now structured, it's "cheaper" to buy a new one. In Egypt and elsewhere in the developing world, whole classes of the poor make their livings by combing through the mountains of rubbish discarded by the better off. To these desperately poor people, even trivial, broken things may be precious, whereas our wealth, leveraged and magnified by technology, cheap energy, and blindness to the ecological costs of throughput, seems too easy. Our appreciation for a thing is divorced from its natural capital cost and no longer in proportion to it.

Merely asking people individually to carry the burden of scaling back is unfair and unrealistic. But if we were to implement natural capital depletion taxes and the other instruments so as to tax throughput rather than income, the market would automatically help to limit the scale of the economy. Prices would be higher, but so would incomes. That combination would encourage saving rather than spending, thus limiting throughput and increasing investment. It would collectively encourage people to pursue development rather than growth.

A properly restructured economic system would help make it easier to consider the proposition that wealth ultimately cannot fulfill us and is a poor substitute for developmental improvements: deeper human relationships, artistic creativity and experience, amateur sports that one actually takes part in, a strong sense of connection to the Earth through the places where we live, and a hundred other forms of human personal enrichment. It would be easier — perhaps even preferable — to consider having less private wealth. If endless growth is impossible, sooner or later we will have to reassess and redefine our material aspirations. From the standpoint of sustainability, sooner would be better. The opportunity simultaneously created is to pursue more meaningful lives.

4. *The value of taking our measure.* It is not hard in principle to understand how the average American, Swiss, or Japanese imposes a greater load on natural capital and the global ecosystem than the average Honduran, Malaysian, or Ghanaian. By every measure, people in the developed world extract more, use more, and waste more. But exactly how much? Finding out can be a consciousness-raising process. It can also reveal important areas for personal conservation and help identify measures that can be taken without excessive sacrifice or inconvenience. (See the Appendix for means to take this kind of inventory.)

A market that was properly adjusted to reflect the real prices of the items and commodities we consume would encourage us to take our measure as consumers — if by no other means than the shock of discovering the real costs of our lifestyles.

5. *The value of full engagement.* We have considerable power as consumers to influence the course of economic life in our communities, our nations, and the world. The billions of dollars spent every year on advertising are one measure of that power; something we have — the power to favor some products or services over others, or to favor none at all — is sought very desperately by corporations. Obviously we should use our power as consumers thoughtfully and for the good as best we can understand it.

That said, it is striking how passively and unconsciously we allow ourselves to be defined as "consumers," as if that were our first and most important purpose, as if we live to consume and not the other way around. Any look at how individuals can help move society toward sustainability must begin by listing some of the other roles we can play: parent, voter, activist, citizen, community member, etc. It is in these roles that the collective decisions of energized individuals can make things happen.

A restructured economic system would encourage deeper investment in other roles by properly making rampant consumerism more costly. At the same time, we should not forget the value of personal engagement. T.S. Eliot noted, in another context, the idea of the "system so perfect that no one needs to be good." Such a system is impossible. No web of laws and regulations will work for long unless people want it to, unless the fundamental sentiments of the people are congruent with the law. And the more people desire and work for the intended ends of the laws, the less the laws are necessary. The political will to implement any of the proposals discussed in this book must come from the bottom up. Conversely, policies imposed from the top down and in the absence of supporting popular sentiment will fail. What is important, then, is the cultivation of the popular sentiment, which begins in individual hearts. Personal belief, commitment, and action are not merely the engines of change but the only sources of integrity. In Gandhi's plain words:

We must *be* the change we wish to see in the world.

Appendix

Some Tools for Personal and Community Action

For reflection:

> *How Much Is Enough? The Consumer Society and the Future of the Earth*, by Alan Durning, Worldwatch Institute/W.W. Norton, New York, 1992.

> "How Earth Friendly Are You? A Lifestyle Self-Assessment Questionnaire." From the Simple Living Network, 1-800-318-5725; URL http://www.simpleliving.net.

For action:

> *50 Simple Things You Can Do To Save the Earth*, from the Natural Resources Defense Council. In bookstores, or contact NRDC, 50 Simple Things, 40 West 20th Street, New York, NY   10011

> "How Green Are You?" Article describing a relatively simple quantitative method of estimating your environmental impact, *Consumer Reports*, p. 725, November, 1994.

A more detailed quantitative assessment of personal or household environmental impact can be performed with EarthAware: Personal Environmental Impact Assessment Software, available in IBM and Macintosh formats from EnviroAccount Software, 605 Sunset Court, Davis, CA 95616; Telephone 916-756-9156.

How Big Is Our Ecological Footprint? A Handbook for Estimating a Community's Appropriated Carrying Capacity (Wackernagel et al., 1993). From the Task Force on Planning Healthy and Sustainable Communities, The University of British Columbia, Department of Family Practice, 5804 Fairview Avenue, Vancouver, BC, Canada V6T 1Z3.

Setting the Stage for Sustainability: A Citizen's Handbook, by Chris Maser, Russ Beaton, and Kevin Smith, Lewis Publishers, Boca Raton, FL, 1998.

"Network to Reduce Overconsumption," a directory of individuals and groups working to promote sustainable consumption, is available from the Center for a New American Dream, 6930 Carroll Avenue, Takoma Park, MD, 20912; telephone 301-891-3683.

Glossary

Allocation. The process and patterns of apportioning extracted or harvested natural resources to various economic uses.

Atomism. As used here, the idea that society is merely the sum of its individual members and that human beings can be seen as economic actors behaving independently from their fellows, rather than as members of communities that influence economic behavior through noneconomic ties and forces.

Biodiversity. The variety of organisms, their functions, and distribution in time and space.

Capital. In this book, capital is classified in three ways. *Natural capital* consists of renewable and nonrenewable resources. *Manufactured (reproducible) capital* consists of the machines, buildings, tools, etc. made by humans from natural capital. *Human capital* includes people's labor, skills, knowledge, and culture.

Carrying capacity. The maximum number of living creatures a territory can support over time without environmental damage.

Complementarity. See substitutability and complementarity.

Discounting. The statistical process by which the effects of uncertainty and interest rates are reflected in calculations of the value of future gains and losses.

Distribution. The process and patterns of dividing up the fruits of economic production among different groups and individuals.

Ecosphere. The zone of the Earth that contains living organisms.

Ecosystem. A biological community and the physical environment associated with it.

Entropy. As used in this book, the tendency of matter and energy to degrade or disperse into less useful forms during economic activity.

Externality. An unintended cost or benefit of production or consumption that is not reflected in the price of the related transactions. Externalities are often borne by people who are not parties to the transactions that create them.

Flows. See stocks and flows.

Primary production. The production of living tissue (biomass) by organisms (such as plants and phytoplankton) that are capable of direct utilization of sunlight via photosynthesis. Primary producers form the foundation of the food web.

Scale. In this book, the size of the human economy relative to the supporting global ecosystem.

Stocks and flows. A stock of a natural resource is the total available quantity of that resource. Stocks of renewable resources, if left alone, will grow. Stocks of nonrenewable resources are essentially fixed. Flows are both the portions of the stocks that are extracted or harvested for economic use and the ecological services rendered by the stocks. If the amount extracted from a renewable resource stock does not exceed the stock's growth rate, then that flow rate is basically perpetual and can be sustained indefinitely. Flows taken from nonrenewable resources always result in diminished stocks.

Substitutability and complementarity. The extent to which one form of capital can be replaced with another to effect production. For example, in recent years automotive engineers have substituted better design (the product of human capital) for natural capital (steel, glass, rubber) to help achieve smaller and lighter vehicles that perform as well or better than their predecessors. Ecological economics argues that natural capital and manufactured capital are largely complements, rather than substitutes.

Throughput. The total volume of matter/energy flowing from the environment through the economy and back into the environment as wastes.

Value. The *use value* of natural capital means the value that derives from its direct use or consumption. *Option value* refers to the value of potential uses, including those made of it by future generations. *Existence value* includes any value unrelated to any current or potential uses.

Welfare. Loosely, human well-being.

References

Acheson, J., (1989), Where have all the exploiters gone? Co-management of the Maine lobster industry, *Common Property Resources: Ecology and Community-Based Sustainable Development*, Berkes, F., Ed., Belhaven Press, London.

Ayres, R.U., (1993), Cowboys, cornucopians and long-run sustainability, *Ecological Economics* 8(3), 189, December.

Ayres, R.U., (1998), Eco-thermodynamics: economics and the second law, *Ecological Economics* 26(2), 189, August.

Barbier, E.B., (1994), Natural capital and the economics of environment and development, *Investing in Natural Capital: The Ecologicial Economics Approach to Sustainability*, Jansson, A., Hammer, M., Folke, C., and Costanza, R., Eds., Island Press, Washington, D.C.

Bauer, H., (1992), *Scientific Literacy and the Myth of the Scientific Method*, University of Illinois Press, Urbana, IL.

Beckerman, W., (1991), National income, *The New Palgrave: The World of Economics*, Eatwell, J., Milgate, M., and Newman, P., Eds., W.W. Norton, New York.

Bergsten, C.F., (1993), The rationale for a rosy view, *The Economist,* 328(7828), 57, September 11.

Berkes, F. and Farvar, M. (1989), Introduction and overview, *Common Property Resources: Ecology and Community-Based Sustainable Development*, Berkes, F., Ed., Belhaven Press, London.

Berkes, F. and Folke, C., (1994), Investing in cultural capital for sustainable use of natural capital, *Investing in Natural Capital: The Ecologicial Economics Approach to Sustainability*, Jansson, A., Hammer, M., Folke, C., and Costanza, R., Eds., Island Press, Washington, D.C.

Berry, W., (1992), Conservation is good work, *The Amicus Journal,* 33, Winter.

Black, R., (1988), Thomas Robert Malthus (1766-1834), *A Lexicon of Economics,* Deane, P. and Kuper, J., Eds., Routledge, London.

Blau, F., (1991), Gender, *The New Palgrave: The World of Economics,* Eatwell, J., Milgate, M., and Newman, P., Eds., W.W. Norton, New York.

167

Boland, L., (1988), Neoclassical economics, *A Lexicon of Economics,* Deane, P. and Kuper, J., Eds., Routledge, London.

Boulding, K., (1964), *The Meaning of the Twentieth Century: The Great Transition,* Harper and Row, New York.

Boulding, K., (1981a), *Ecodynamics*. Sage Publications, Beverly Hills.

Boulding, K., (1981b), *Evolutionary Economics,* Sage Publications, Beverly Hills.

Clark, C., (1989), Clear-cut economies: Should we harvest everything now? *The Sciences,* 29(1), January 1.

Cleveland, C., (1994), Re-allocating work between human and natural capital in agriculture: Examples from India and the United States, *Investing in Natural Capital: The Ecologicial Economics Approach to Sustainability,* Jansson, A., Hammer, M., Folke, C., and Costanza, R., Eds., Island Press, Washington, D.C.

Costanza, R., (1987), Social traps and environmental policy, *BioScience,* 37(6), 407.

Costanza, R., (1994), Three general policies to achieve sustainability, *Investing in Natural Capital: The Ecologicial Economics Approach to Sustainability,* Jansson, A., Hammer, M., Folke, C., and Costanza, R., Eds., Island Press,Washington, D.C.

Costanza, R. and Daly, H.E., (1992), Natural capital and sustainable development, *Conservation Biology,* 6(1), 37.

Costanza, R., Kemp, W., and Boynton, W., (1993), Predictability, scale and biodiversity in coastal and estuarine ecosystems: Implications for management, *Ambio,* 22(2-3), 88, May.

Costanza, R., Cumberland, J., Daly, H., Goodland, R., and Norgaard, R., (1997), *An Introduction to Ecological Economics,* St. Lucie Press, Boca Raton, FL.

Daly, H.E., (1991a), Elements of environmental macroeconomics, *Ecological Economics: The Science and Management of Sustainability,* Costanza, R., Ed., Columbia University Press, New York.

Daly, H.E., (1991b), *Steady-State Economics*, 2nd ed., Island Press, Washington, D.C.

Daly, H.E., (1992), Steady-state economics: Concepts, questions, policies, *GAIA,* 6, 333.

Daly, H.E., (1993), The perils of free trade, *Scientific American,* 269(5), 24, November.

Daly, H.E., (1994), *Beyond Growth: The Economics of Sustainable Development,* Beacon Press, Boston.

Daly, H.E. and Cobb, J., (1989), *For the Common Good: Redirecting the Economy Toward Community, the Environment, and a Sustainable Future,* Beacon Press, Boston.

d'Arge, R.C., (1994), Sustenance and sustainability: How can we preserve and consume without major conflict? *Investing in Natural Capital: The Ecologicial Economics Approach to Sustainability,* Jansson, A., Hammer, M., Folke, C., and Costanza, R., Eds., Island Press, Washington, D.C.

Daviss, B., (1998), Let's get emotional, *New Scientist,* 39, September 19.

De Groot, R., (1994), Environmental functions and the economic value of eco-systems, draft chapter for *Investing in Natural Capital: The Ecologicial Economics Approach to Sustainability*, Jansson, A., Hammer, M., Folke, C., and Costanza, R., Eds., Island Press, Washington, D.C.

Easterlin, R., (1974), Does economic growth improve the human lot? Some empirical evidence, *Nations and Households in Economic Growth: Essays in Honor of Moses Abramovitz*, David, P. and Reder, M., Eds., Academic Press, New York.

Ehrlich, A. and Ehrlich, P., (1986), Population and development misunderstood., *The Amicus Journal*, 8, Summer.

Ehrlich, P. and Ehrlich, A., (1991), The value of biodiversity, unpublished manuscript.

Ekins, P., Folke, C., and Costanza, R., (1994), Trade, environment and development: the issues in perspective, *Ecological Economics*, 9, 1, January.

Energy Information Administration, (1993), *Emissions of Greenhouse Gases in the United States 1985-1990*, DOE/EIA-0573, U.S. Department of Energy, Washington, D.C.

Engelman, R. and LeRoy, P., (1993), *Sustaining Water: Population and the Future of Renewable Water Supplies*, Population Action International, Washington, D.C.

Findlay, R., (1991), Comparative advantage, *The New Palgrave: The World of Economics*, Eatwell, J., Milgate, M., and Newman, P., Eds., W.W. Norton, New York.

Fremlin, J.H., (1964), How many people can the world support?, *New Scientist*, 29, 285, October.

Gever, J., Kaufmann, R., Skole, D., and Vorosmarty, C., (1991), *Beyond Oil: The Threat to Food and Fuel in the Coming Decades*, University of Colorado Press, Niwot, CO.

Goodland, R., Ledec, G., and Webb, M., (1989), Meeting environmental concerns caused by common-property mismanagement in economic development projects, *Common Property Resources: Ecology and Community-Based Sustainable Development*, Berkes, F., Ed., Belhaven Press, London.

Gowdy, J., Ed., (1998), *Limited Wants, Unlimited Means: A Reader on Hunter-Gatherer Economics and the Environment*, Island Press, Washington, D.C.

Grant, L., (1992), *Elephants in the Volkswagen*, W.H. Freeman and Company, New York, cited in *Focus*, 3(2), 31, 1993.

Grima, A. and Berkes, F., (1989), Natural resources: Access, rights-to-use and management, *Common Property Resources: Ecology and Community-Based Sustainable Development*, Berkes, F., Ed., Belhaven Press, London.

Hanna, S., (1994), Linking human and natural systems through property rights regimes, paper presented at the third biannual conference of the International Society for Ecological Economics, San Jose, Costa Rica, October 24-28.

Hardin, G., (1968), The tragedy of the commons, *Science*, 162, 1243, December 13.

Hardin, G., (1986), Cultural carrying capacity: A biological approach to human problems, *BioScience*, 36, 599.

Hawken, P., (1993), *The Ecology of Commerce*, HarperBusiness, New York.

Henderson, H., (1981), *The Politics of the Solar Age: Alternatives to Economics*, Anchor Press, New York.

Holling, C.S., (1973), Resilience and stability of ecological systems, *Annual Review of Ecology and Systematics*, 4, 1.

Huston, M., (1993), Biological diversity, soils, and economics, *Science*, 262, 1676, December 10.

Jackson, W., (1991), Nature as the measure for sustainable agriculture, *Ecology, Economics, Ethics: The Broken Circle*, Bormann, F. and Kellert, S., Eds., Yale University Press, New Haven.

Jansson, A. and Jannson, B., (1994), Ecosystem properties as basis for sustainability. *Investing in Natural Capital: The Ecologicial Economics Approach to Sustainability*, Jansson, A., Hammer, M., Folke, C., and Costanza, R., Eds., Island Press, Washington, D.C.

Janzen, D., (1998), Personal communication, September 25.

King, D., (1994), Can we justify sustainability? New challenges facing ecological economics, *Investing in Natural Capital: The Ecologicial Economics Approach to Sustainability*, Jansson, A., Hammer, M., Folke, C., and Costanza, R., Eds., Island Press, Washington, D.C.

Krugman, P., (1990), *The Age of Diminished Expectations: U.S. Economic Policy in the 1990s*, MIT Press, Cambridge, MA.

Leontief, W., (1982), Academic economics, *Science*, 217, 104, July 9.

Lewis, P., (1993), New U.N. index measures wealth as quality of life, *The New York Times*, p. 14, May 23.

Lovins, A., (1998), "On the rebound" (letter), *New Scientist*, 52, October 10.

MacKenzie, J., Dower, R., and Chen, D., (1992), *The Going Rate: What It Really Costs To Drive*, World Resources Institute, Washington, D.C.

Maler, K., Gren, I., and Folke, C., (1994), Multiple use of environmental resources: A household production function approach to valuing natural capital, *Investing in Natural Capital: The Ecological Economics Approach to Sustainability*, Jansson, A., Hammer, M., Folke, C., and Costanza, R., Eds., Island Press, Washington, D.C.

McFarling, U., (1994), New report predicts extinction of many species, *The Boston Globe*, 28, September 7.

Mill, J.S., (1969) (1848), *Principles of Political Economy, with Some of Their Applications to Social Philosophy*, Augustus M. Kelley, New York.

Naeem, S., Thompson, L., Lawler, S., Lawton, J., and Woodfin, R., (1994), Declining biodiversity can alter the performance of ecosystems, *Nature*, 368, 734, April 21.

Nash, R., (1982), *Wilderness and the American Mind*, Yale University Press, New Haven, CT.

Norgaard, R., (1994), *Development Betrayed: The End of Progress and a Coevolutionary Revisioning of the Future*, Routledge, New York.

Passell, P., (1993), Disputed new role for polls: Putting a price tag on nature, *The New York Times*, September 6.

Pearce, D. and Turner, R., (1990), *Economics of Natural Resources and the Environment*, The Johns Hopkins University Press, Baltimore.

Pearce, F., (1998), Consuming myths, *New Scientist*, 5, 18, September.

Perrings, C., (1994), Biotic diversity, sustainable development and natural capital, *Investing in Natural Capital: The Ecological Economics Approach to Sustainability*, Jansson, A., Hammer, M., Folke, C., and Costanza, R., Eds., Island Press, Washington, D.C.

Ponting, C., (1991), *A Green History of the World: The Environment and the Collapse of Great Civilizations*, Penguin Books, New York.

Postel, S., (1989), Halting land degradation, *State of the World 1989*, Brown, L. et al., W.W. Norton, New York.

Prince, R. and Gordon, P. (1994), *Greening the National Accounts*, U.S. Congressional Budget Office, Washington, D.C., March.

Rees, W.E. and Wackernagel, M., (1994), Ecological footprints and appropriated carrying capacity: Measuring the natural capital requirements of the human economy, *Investing in Natural Capital: The Ecological Economics Approach to Sustainability*, Jansson, A., Hammer, M., Folke, C., and Costanza, R., Eds., Island Press, Washington, D.C.

Repetto, R., (1992), Accounting for environmental assets, *Scientific American*, 94, June.

Repetto, R., (1993), What can policymakers learn from natural resource accounting? Paper presented to the Organization of American States Seminar on Natural Resource and Environmental Accounts for Development Policy, Washington, D.C., April 13-14.

Ropke, I., (1994), Trade, development and sustainability — a critical assessment of the "free trade dogma," *Ecological Economics*, 9, 13, January.

Schlefer, J., (1993), History counsels "no" on NAFTA, *The New York Times*, p. F11, November 14.

Schrödinger, E., (1967), *What Is Life?*, Cambridge University Press, London.

Segura Bonilla, O. and Boyce, J., (1994), Investing in natural and human capital in developing countries, *Investing in Natural Capital: The Ecological Economics Approach to Sustainability*, Jansson, A., Hammer, M., Folke, C., and Costanza, R., Eds., Island Press, Washington, D.C.

Skinner, A., (1988), Adam Smith (1723-90), *A Lexicon of Economics*, Deane, P. and Kuper, J., Eds., Routledge, London.

Stiglitz, J., (1986), Discount rates for social cost-benefit analysis, *Environmental Law and Policy: A Coursebook on Nature, Law, and Society*, Plater, Z., Abrams, R., and Goldfarb, W., West Publishing Company, St. Paul, MN, 1992.

Tainter, J., (1988), *The Collapse of Complex Societies*, Cambridge University Press, Cambridge.

Tilman, D., May, R., Lehman, C., and Nowak, M., (1994), Habitat destruction and the extinction debt, *Nature*, 371, 65, September 1.

Vitousek, P., Ehrlich, P., Ehrlich, A., and Matson, P., (1986), Human appropriation of the products of photosynthesis, *BioScience*, 36(6), 368.

Waldrop, M. Mitchell, (1992), *Complexity: The Emerging Science at the Edge of Order and Chaos*, Simon and Schuster, New York.

White, L. , (1967), The historical roots of our ecological crisis, *Science*, 155(3767), 1203, March 10.

Wilson, E.O., (1992), *The Diversity of Life,* Belknap Harvard, Cambridge, MA.

World Resources Institute, (1992), *World Resources 1992-93: A Guide to the Global Environment,* Oxford University Press, New York.

World Resources Institute, (1994), *World Resources 1994-95: A Guide to the Global Environment,* Oxford University Press, New York.

World Resources Institute, (1998), *World Resources 1998-99: A Guide to the Global Environment,* Oxford University Press, New York.

Index

173